PROBLEM-SOLVING CASES IN MICROSOFT® ACCESS AND EXCEL,
SIXTH ANNUAL EDITION

Joseph A. Brady

Ellen F. Monk

THOMSON

COURSE TECHNOLOGY

Australia • Canada • Mexico • Singapore • Spain • United Kingdom • United States

THOMSON

COURSE TECHNOLOGY ™

Problem-Solving Cases in Microsoft® Access and Excel,
Sixth Annual Edition

Product Manager:
Kate Hennessy

Development Editor:
Dan Seiter

Marketing Manager:
Bryant Chrzan

Marketing Specialist:
Vicki Ortiz

Editorial Assistant:
Patrick Frank

Content Product Manager:
Danielle Chouhan
Matthew Hutchinson

Manufacturing Coordinator:
Justin Palmeiro

Cover Designer:
Marissa Falco

Compositor:
GEX Publishing Services

Disclaimer

Cengage Course Technology reserves the right to revise this publication and make changes from time to time in its content without notice.

Cengage Course Technology and the Cengage Course Technology logo are registered trademarks used under license. All other names and visual representations used herein are for identification purposes only and are trademarks and the property of their respective owners.

ISBN 13: 978-1-4239-0213-3
ISBN 10: 1-4239-0213-0

Dedication

To Sparky—sometimes a problem, always a joy
JAB
To my husband, Peter, who encourages me in my academic pursuits
EFM

TABLE OF CONTENTS

For nearly two decades, we have taught MIS courses at the University of Delaware. From the start, we wanted to use good computer-based case studies for the database and decision-support portions of our courses.

We could not find a casebook that met our needs! This surprised us because our requirements, we thought, were not unreasonable. First, we wanted cases that asked students to think about real-world business situations. Second, we wanted cases that provided students with hands-on experience, using the kind of software that they had learned to use in their computer literacy courses—and that they would later use in business. Third, we wanted cases that would strengthen students' ability to analyze a problem, examine alternative solutions, and implement a solution using software. Undeterred by the lack of casebooks, we wrote our own cases, and Cengage Course Technology published them.

This book is the seventh casebook we have written for Cengage Course Technology. The cases are all new and the tutorials are updated. New features of this textbook reflect comments and suggestions that we have received from instructors who have used our casebook in their classrooms. In addition, the current cases have been written for the new Office 2007 versions of Microsoft Access and Excel.

As with our prior casebooks, we include tutorials that prepare students for the cases, which are challenging but doable. Most of the cases are organized in a way that helps students think about the logic of each case's business problem and then about how to use the software to solve the business problem. The cases fit well in an undergraduate MIS course, an MBA Information Systems course, or a Computer Science course devoted to business-oriented programming.

BOOK ORGANIZATION

The book is organized into seven parts:

1. Database Cases Using Access

2. Decision Support Cases Using Excel Scenario Manager

3. Decision Support Cases Using the Excel Solver

4. Decision Support Case Using Basic Excel Functionality

5. Integration Cases: Using Access and Excel

6. Advanced Excel Skills Using Excel

7. Presentation Skills

Part 1 begins with two tutorials that prepare students for the Access case studies. Parts 2 and 3 each begin with a tutorial that prepares students for the Excel case studies. All four tutorials provide students with hands-on practice in using the software's more advanced features—the kind of support that other books about Access and Excel do not give to students. Part 4 asks students to use Excel's basic functionality for decision support. Part 5 challenges students to use both Access and Excel to find a solution to a business problem. Part 6 is a set of short tutorials on the advanced skills that students need to complete some of the Excel cases. Part 7 is a tutorial that hones students' skills in creating and delivering an oral presentation to business managers. The next section explores each of these parts in more depth.

Part 1: Database Cases
Using Access
This section begins with two tutorials and then presents five case studies.

Tutorial A: Database Design
This tutorial helps students to understand how to set up tables to create a database, without requiring students to learn formal analysis and design methods, such as data normalization.

Tutorial B: Microsoft Access

The second tutorial teaches students the more advanced features of Access queries and reports—features that students will need to know to complete the cases.

Cases 1–5

Five database cases follow Tutorials A and B. The students' job is to implement each case's database in Access so form, query, navigation pane, and report outputs can help management. The first case is an easier "warm-up" case. The next four cases require a more demanding database design and implementation effort.

Part 2: Decision Support Cases
Using the Excel Scenario Manager

This section has one tutorial and two decision support cases requiring the use of Excel Scenario Manager.

Tutorial C: Building a Decision Support System in Excel

This section begins with a tutorial using Excel for decision support and spreadsheet design. Fundamental spreadsheet design concepts are taught. Instruction on the Scenario Manager, which can be used to organize the output of many "what-if" scenarios, is emphasized.

Cases 6–7

These two cases can be done with or without the Scenario Manager (although the Scenario Manager is nicely suited to them). In each case, students must use Excel to model two or more solutions to a problem. Students then use the outputs of the model to identify and document the preferred solution via a memorandum and, if assigned to do so, an oral presentation.

Part 3: Decision Support Cases
Using the Excel Solver

This section has one tutorial and two decision support cases requiring the use of Excel Solver.

Tutorial D: Building a Decision Support System Using Excel Solver

This section begins with a tutorial about using the Solver, which is a decision support tool for solving optimization problems.

Cases 8–9

Once again, in each case, students use Excel to analyze alternatives and identify and document the preferred solution.

Part 4: Decision Support Case
Using Basic Excel Functionality

Case 10

The cases continue with one case that uses basic Excel functionality (in other words, the case does not require the Scenario Manager or the Solver). Excel is used to test students' analytical skills in "what if" analyses.

Part 5: Integration Cases
Using Excel and Access

Cases 11 and 12

These cases integrate Access and Excel. These cases are included because of a trend toward sharing data among multiple software packages to solve problems.

Part 6: Advanced Excel Skills Using Excel

Using Excel

This part has one tutorial focused on advanced techniques in Excel.

Tutorial E: Guidance for Excel Cases

A number of the cases in this book require the use of some advanced techniques in Excel. The following techniques are explained in this tutorial (rather than in the cases themselves): Data tables and Pivot tables.

Part 7: Presentation Skills

Tutorial F: Giving an Oral Presentation

Each case includes an optional presentation assignment that gives students practice in making a presentation to management on the results of their analysis of the case. This section gives advice on how to create oral presentations. It also has technical information on charting, a technique that might be useful in case analyses or as support for presentations. This tutorial will help students to organize their recommendations, to present their solutions in both words and graphics, and to answer questions from the audience. For larger classes, instructors may wish to have students work in teams to create and deliver their presentations—which would model the "team" approach used by many corporations.

To view and access additional cases, instructors should see the "Hall of Fame" note in the *Using the Cases* section shown below.

INDIVIDUAL CASE DESIGN

The format of the cases follows this template.

- Each case begins with a *Preview* of what the case is about and an overview of the tasks.

- The next section, *Preparation*, tells students what they need to do or know to complete the case successfully. (Of course, our tutorials prepare students for the cases!)

- The third section, *Background*, provides the business context that frames the case. The background of each case models situations that require the kinds of thinking and analysis that students will need in the business world.

- The Background sections are followed by the *Assignment* sections, which are generally organized in a way that helps students to develop their analyses.

- The last section, *Deliverables*, lists what students must hand in: printouts, a memorandum, a presentation, and files on disk. The list is similar to the kind of deliverables that a business manager might demand.

USING THE CASES

We have successfully used cases like these in our undergraduate MIS courses. We usually begin the semester with Access database instruction. We assign the Access database tutorials and then a case to each student. Then, for Excel DSS instruction, we do the same thing—assign a tutorial and then a case.

Some instructors have expressed an interest in having access to extra cases, especially in the second semester of a school year. For example: "I assigned the integration case in the fall, and need another one for the spring." To meet this need, we have set up an online "Hall of Fame" that features some of our favorite cases from prior editions. This password-protected Hall of Fame is available to instructors and can be found on the Cengage Course Technology Web site. Go to *www.course.com* and search for this textbook by title, author, or ISBN. Note that the cases are in MS Office 2003 format, but MS Office 2007 will read and tranlate them easily.

TECHNICAL INFORMATION

This textbook was quality assurance tested using the Windows XP Professional operating system, Microsoft Access 2007, and Microsoft Excel 2007.

Data Files and Solution Files

We have created "starter" data files for the Excel cases, so students need not spend time typing in the spreadsheet skeleton. Case 11 also requires students to load an Access database file. All these files can be found on the Cengage Course Technology Web site, which is available to both students and instructors. Go to *www.course.com* and search for this textbook by title, author, or ISBN. You are granted a license to copy the data files to any computer or computer network used by individuals who have purchased this textbook.

Solutions to the material in the text are available to instructors. These can also be found at *www.course.com*. Search for this textbook by title, author, or ISBN. The solutions are password protected.

Instructor's Manual

An Instructor's Manual is available to accompany this text. The Instructor's Manual contains additional tools and information to help instructors successfully use this textbook. Items such as a Sample Syllabus, Teaching Tips, and Grading Guidelines can be found in the Instructor's Manual. Instructors should go to *www.course.com* and search for this textbook by title, author, or ISBN. The Instructor's Manual is password protected.

ACKNOWLEDGEMENTS

We would like to give many thanks to the team at Cengage Course Technology, including our Developmental Editor, Dan Seiter; Product Manager, Kate Hennessy; and our Content Project Managers, Danielle Chouhan and Matt Hutchinson. As always, we acknowledge our students' diligent work.

DATABASE CASES
USING ACCESS

DATABASE DESIGN

This tutorial has three sections. The first section briefly reviews basic database terminology. The second section teaches database design. The third section features a practice database design problem.

REVIEW OF TERMINOLOGY

You will begin by reviewing some basic terms that will be used throughout this textbook. In Access, a **database** is a group of related objects that are saved in one file. An Access **object** can be a table, a form, a query, or a report. You can identify an Access database file by its suffix .accdb.

A **table** consists of data that is arrayed in rows and columns. A **row** of data is called a **record**. A **column** of data is called a **field**. Thus, a record is a set of related fields. The fields in a table should be related to one another in some way. For example, a company might want to keep its employee data together by creating a database table called EMPLOYEE. That table would contain data fields about employees, such as their names and addresses. It would not have data fields about the company's customers; that data would go in a CUSTOMER table.

A field's values have a **data type** that is declared when a table is defined. That way, when data is entered into the database, the software knows how to interpret each entry. The data types found in Access include the following:

- "Text" for words
- "Integer" for whole numbers
- "Double" for numbers that have a decimal value
- "Currency" for numbers that should be treated as dollars and cents
- "Yes/No" for variables that have only two values (1-0, on/off, yes/no, true/false)
- "Date/Time" for variables that are dates or times

Each database table should have a **primary key** field, a field in which each record has a *unique* value. For example, in an EMPLOYEE table, a field called SSN (for Social Security number) could serve as a primary key, because each record's SSN value would be different from every other record's SSN value. Sometimes a table does not have a single field whose values are all different. In that case, two or more fields are combined into a **compound primary key**. The combination of the fields' values is unique.

Database tables should be logically related to one another. For example, suppose a company has an EMPLOYEE table with fields for SSN, Name, Address, and Telephone Number. For payroll purposes, the company has an HOURS WORKED table with a field that summarizes Labor Hours for individual employees. The relationship between the EMPLOYEE table and the HOURS WORKED table needs to be established in the database so you can tell which employees worked which hours. That is done by including the primary key field from the EMPLOYEE table (SSN) as a field in the HOURS WORKED table. In the HOURS WORKED table, the SSN field is then called a **foreign key**.

In Access, data can be entered directly into a table or it can be entered into a form, which then inserts the data into a table. A **form** is a database object that is created from an existing table to make the process of entering data more user-friendly.

A **query** is the database equivalent of a question that is posed about data in a table (or tables). For example, suppose a manager wants to know the names of employees who have worked for the company for more than five years. A query could be designed in such a way that it interrogates the EMPLOYEE table to search for the information. The query would be run, and its output would answer the question.

Because a query may need to pull data from more than one table, queries can be designed to interrogate more than one table at a time. In that case, the tables must be connected by a **join** operation, which links tables on the values in a field that they have in common. The common field acts as a "hinge" for the joined tables; when the query is run, the query generator treats the joined tables as one large table.

In Access, queries that answer a question are called select queries, because they select relevant data from the database records. Queries also can be designed to change data in records, add a record to the end of a table, or delete entire records from a table. Those are called **update**, **append**, and **delete** queries, respectively.

Access has a **report** generator that can be used to format a table's data or a query's output.

DATABASE DESIGN

Designing a database involves determining which tables need to be in the database and creating the fields that need to be in each table. This section begins with an introduction to key database design concepts, then discusses database design rules. Those rules are a series of steps you should use when building a database. First, the following key concepts are defined:

- Entities
- Relationships
- Attributes

Database Design Concepts

Computer scientists have highly formalized ways of documenting a database's logic, but learning their notations and mechanics can be time-consuming and difficult. In fact, doing so usually takes a good portion of a systems analysis and design course. This tutorial will teach you database design by emphasizing practical business knowledge, and the approach should enable you to design serviceable databases. Your instructor may add more formal techniques.

A database models the logic of an organization's operation, so your first task is to understand that operation. You do that by talking to managers and workers, by making observations, and/or by looking at business documents such as sales records. Your goal is to identify the business's "entities" (sometimes called *objects*). An **entity** is some thing or some event that the database will contain. Every entity has characteristics, called **attributes**, and a **relationship** (or relationships) to other entities. Take a closer look.

Entities

As previously mentioned, an entity is a tangible thing or an event. The reason for identifying entities is that *an entity eventually becomes a table in the database*. Entities that are things are easy to identify. For example, consider a video store. The database for the video store would probably need to contain the names of DVDs and the names of customers who rent them, so you would have one entity named VIDEO and another named CUSTOMER.

In contrast, entities that are events can be more difficult to identify, probably because although events cannot be seen, they are no less real. In the video store example, one event would be VIDEO RENTAL and another event would be HOURS WORKED by employees.

In general, your analysis of an organization's operations can be made easier by knowing that organizations usually have certain physical entities such as these:

- Employees
- Customers
- Inventory (products or services)
- Suppliers

Thus, the database for most organizations would have a table for each of those entities. Your analysis also can be made easier by knowing that organizations engage in transactions internally (within the company) and externally (with the outside world). Those transactions are the subject of any accounting course, but most people understand them from events that occur in daily life. Consider the following examples:

- Organizations generate revenue from sales or interest earned. Revenue-generating transactions include event entities called SALES and INTEREST.
- Organizations incur expenses from paying hourly employees and purchasing materials from suppliers. HOURS WORKED and PURCHASES are event entities in the databases of most organizations.

Thus, identifying entities is a matter of observing what happens in an organization. Your powers of observation are aided by knowing what entities exist in the databases of most organizations.

Relationships

As a database analyst building a database, you should consider the relationship of each entity to other entities. For each entity, you should ask, "What is the relationship, if any, of this entity to every other entity identified?" Relationships can be expressed in English. For example, suppose a college's database has entities for STUDENT (containing data about each student), COURSE (containing data about each course), and SECTION (containing data about each section). A relationship between STUDENT and SECTION would be expressed as "Students enroll in sections."

An analyst also must consider the **cardinality** of any relationship. Cardinality can be one-to-one, one-to-many, or many-to-many. Those relationships are summarized as follows:

- In a one-to-one relationship, one instance of the first entity is related to just one instance of the second entity.
- In a one-to-many relationship, one instance of the first entity is related to many instances of the second entity, but each instance of the second entity is related to only one instance of the first entity.
- In a many-to-many relationship, one instance of the first entity is related to many instances of the second entity and one instance of the second entity is related to many instances of the first entity.

For a more concrete understanding of cardinality, consider again the college database with the STUDENT, COURSE, and SECTION entities. The university catalog shows that a course such as Accounting 101 can have more than one section: 01, 02, 03, 04, etc. Thus, the following relationships can be observed:

- The relationship between the entities COURSE and SECTION is one-to-many. Each course has many sections, but each section is associated with just one course.
- The relationship between STUDENT and SECTION is many-to-many. Each student can be in more than one section, because each student can take more than one course. Also, each section has more than one student.

Thinking about relationships and their cardinalities may seem tedious to you. But as you work through the cases in this text, you will see that this type of analysis and the knowledge it yields can be very valuable in designing databases. In the case of many-to-many relationships, you should determine the database tables a given database needs; in the case of one-to-many relationships, you should decide which fields the database's tables need to share.

Attributes

An attribute is a characteristic of an entity. The reason you identify attributes of an entity is because *attributes become a table's fields*. If an entity can be thought of as a noun, an attribute can be thought of as an adjective describing the noun. Continuing with the college database example, consider the STUDENT entity. Students have names. Thus, Last Name would be an attribute of the entity called STUDENT and, therefore, a field in the STUDENT table. First Name would be an attribute as well. The STUDENT entity also would have an Address attribute as another field, along with Phone Number and whatever other fields were used.

Sometimes it can be difficult to tell the difference between an attribute and an entity. One good way to differentiate them is to ask whether more than one attribute is possible for each entity. If more than one instance is possible and you do not know in advance how many there will be, then it's an entity. For example, assume a student could have two (but no more than two) Addresses—one for home and one for college. You could specify attributes Address 1 and Address 2. Now consider what would happen if the number of student addresses could not be stipulated in advance, meaning all addresses had to be recorded. In that case, you would not know how many fields to set aside in the STUDENT table for addresses. Therefore, you would need a separate STUDENT ADDRESSES table (entity) that would show any number of addresses for a given student.

DATABASE DESIGN RULES

As described previously, your first task in database design is to understand the logic of the business situation. Once you understand that, you are ready to build a database for the requirements of the situation. To create a context for learning about database design, look at a hypothetical business operation and its database needs.

Example: The Talent Agency

Suppose you have been asked to build a database for a talent agency. The agency books bands into nightclubs. The agent needs a database to keep track of the agency's transactions and to answer day-to-day questions. Many questions arise in the running of this business. For example, a club manager often wants to know which bands are available on a certain date at a certain time or wants to know the agent's fee for a certain band. Similarly, the agent may want to see a list of all band members and the instrument each person plays or a list of all bands having three members.

Suppose you have talked to the agent and have observed the agency's business operation. You conclude that your database needs to reflect the following facts:

1. A booking is an event in which a certain band plays in a particular club on a particular date, starting at a certain time, ending at a certain time, and performing for a specific fee. A band can play more than once a day. The Heartbreakers, for example, could play at the East End Cafe in the afternoon and then at the West End Cafe that same night. For each booking, the club pays the talent agent. The agent keeps a 5 percent fee and then gives the remainder of the payment to the band.

2. Each band has at least two members and an unlimited maximum number of members. The agent notes a telephone number of just one band member, which is used as the band's contact number. No two bands have the same name or telephone number.

3. No members of any of the bands have the same name. For example, if there is a Sally Smith in one band, there is no Sally Smith in another band.

4. The agent keeps track of just one instrument that each band member plays. For this record keeping purpose, "vocals" are considered an instrument.

5. Each band has a desired fee. For example, the Lightmetal band might want $700 per booking and would expect the agent to try to get at least that amount.

6. Each nightclub has a name, an address, and a contact person. That person has a telephone number that the agent uses to contact the club. No two clubs have the same name, contact person, or telephone number. Each club has a target fee. The contact person will try to get the agent to accept that amount for a band's appearance.

7. Some clubs feed the band members for free; others do not.

Before continuing with this tutorial, you might try to design the agency's database on your own. Ask yourself, What are the entities? Recall that databases usually have CUSTOMER, EMPLOYEE, and INVENTORY entities as well as an entity for the revenue-generating transaction event. Each entity becomes a table in the database. What are the relationships between entities? For each entity, what are its attributes? For each table, what is the primary key?

Six Database Design Rules

Assume you have gathered information about the business situation in the talent agency example. Now you want to identify the tables required for the database and the fields needed in each table. To do that, observe the following six rules:

Rule 1: You do not need a table for the business. The database represents the entire business. Thus, in the example, Agent and Agency are not entities.

Rule 2: Identify the entities in the business description. Look for the things and events that the database must contain. Those become tables in the database. Typically, certain entities are represented. In the talent agency example, you should be able to observe these entities:

*   *Things*: The product (inventory for sale) is Band. The customer is Club.
*   *Events*: The revenue-generating transaction is Bookings.

You might ask yourself, Is there an EMPLOYEE entity? Isn't INSTRUMENT an entity? Those issues will be discussed as the rules are explained.

Rule 3: Look for relationships between the entities. Look for one-to-many relationships between entities. The relationship between those entities must be established in the tables, and that is done by using a foreign key. Those mechanics are explained in the next rule's discussion of the relationship between Band and Band Member.

Look for many-to-many relationships between entities. In each of those relationships, there is the need for a third entity that associates the two entities in the relationship. Recall from the college database scenario the many-to-many relationship example that involved STUDENT and SECTION entities. To show the ENROLLMENT of specific students in specific sections, a third table needs to be created. The mechanics of doing that are described in the next rule in the discussion of the relationship between BAND and CLUB.

Rule 4: Look for attributes of each entity and designate a primary key. As previously mentioned, you should think of the entities in your database as nouns. You should then create a list of adjectives that describe those nouns. Those adjectives are the attributes that will become the table's fields. After you have identified fields for each table, you should check to see if a field has unique values. If one exists, designate it as the primary key field; otherwise, designate a compound primary key.

Returning to the talent agency example, the attributes, or fields, of the BAND entity are Band Name, Band Phone Number, and Desired Fee. No two bands have the same names, so the primary key field can be Band Name. Figure A-1 shows the BAND table and its fields: Band Name, Band Phone Number, and Desired Fee; the data type of each field also is shown.

BAND

Field	Data Type
Band Name (primary key)	Text
Band Phone Number	Text
Desired Fee	Currency

FIGURE A-1 The BAND table and its fields

Two BAND records are shown in Figure A-2.

Band Name (primary key)	Band Phone Number	Desired Fee
Heartbreakers	981 831 1765	$800
Lightmetal	981 831 2000	$700

FIGURE A-2 Records in the BAND table

If two bands could have the same name (that is, if uniqueness in band names wasn't assumed), Band Name would not be a good primary key and some other unique identifier would be needed. Those situations are common. Most businesses have many types of inventory, and duplicate names are possible. The typical solution is to assign a number to each product to be used as the primary key field. For example, a college could have more than one faculty member with the same name, so each faculty member would be assigned an employee identification number (EIN). Similarly, banks assign a personal identification number (PIN) for each depositor. Each automobile that a car manufacturer makes gets a unique Vehicle Identification Number (VIN). Most businesses assign a number to each sale, called an invoice number. (The next time you buy something at a grocery store, note the number on your receipt. It will be different from the number the next person in line sees on his or her receipt.)

At this point, you might be wondering why Band Member would not be an attribute of BAND. The answer is that although you must record each band member, you do not know in advance how many members will be in each band. Therefore, you do not know how many fields to allocate to the BAND table for members. Another way to think about Band Member(s) is that they are, in effect, the agency's employees. Databases for organizations usually have an EMPLOYEE entity. Therefore, you should create a BAND MEMBER table with the attributes Member Name, Band Name, Instrument, and Phone. The BAND MEMBER table and its fields are shown in Figure A-3.

BAND MEMBER	
Field Name	Data Type
Member Name (primary key)	Text
Band Name (foreign key)	Text
Instrument	Text
Phone	Text

FIGURE A-3   The BAND MEMBER table and its fields

Note in Figure A-3 that the phone number is classified as a text data type. The data type for such "numbers" is text—and not number—because the values will not be used in an arithmetic computation. The benefit is that text data type values take up fewer bytes than numerical or currency data type values; therefore, the file uses less storage space. That rule of using text data types would also hold for values such as ZIP Codes, Social Security numbers, etc.

Five records in the BAND MEMBER table are shown in Figure A-4.

Member Name (primary key)	Band Name	Instrument	Phone
Pete Goff	Heartbreakers	Guitar	981 444 1111
Joe Goff	Heartbreakers	Vocals	981 444 1234
Sue Smith	Heartbreakers	Keyboard	981 555 1199
Joe Jackson	Lightmetal	Sax	981 888 1654
Sue Hoopes	Lightmetal	Piano	981 888 1765

FIGURE A-4   Records in the BAND MEMBER table

Instrument can be included as a field in the BAND MEMBER table, because the agent records only one instrument for each band member. Thus, the instrument can be thought of as a way to describe a band member, much like the phone number is part of the description. Member Name can be the primary key because of the assumption (albeit arbitrary) that no two members in any band have the same name. Alternatively, Phone could be the primary key, assuming no two members share a telephone. Or a band member ID number could be assigned to each person in each band, which would create a unique identifier for each band member that the agency handled.

You might ask why Band Name is included in the BAND MEMBER table. The commonsense reason is that you did not include the Member Name in the BAND table. You must relate bands and members somewhere, and the BAND table is the place to do it.

Another way to think about this involves the cardinality of the relationship between BAND and BAND MEMBER. It is a one-to-many relationship: one band has many members, but each member is in just one band. You establish that kind of relationship in the database by using the primary key field of one table as a foreign key in the other table. In BAND MEMBER, the foreign key Band Name is used to establish the relationship between the member and his or her band.

The attributes of the entity CLUB are Club Name, Address, Contact Name, Club Phone Number, Preferred Fee, and Feed Band?. The table called CLUB can define the CLUB entity, as shown in Figure A-5.

CLUB Field Name	Data Type
Club Name (primary key)	Text
Address	Text
Contact Name	Text
Club Phone Number	Text
Preferred Fee	Currency
Feed Band?	Yes/No

FIGURE A-5   The CLUB table and its fields

Two records in the CLUB table are shown in Figure A-6.

Club Name (primary key)	Address	Contact Name	Club Phone Number	Preferred Fee	Feed Band?
East End	1 Duce St.	Al Pots	981 444 8877	$600	Yes
West End	99 Duce St.	Val Dots	981 555 0011	$650	No

FIGURE A-6   Records in the CLUB table

You might wonder why Bands Booked into Club (or some such field name) is not an attribute of the CLUB table. There are two reasons. First, because you do not know in advance how many bookings a club will have, the value cannot be an attribute. Second, BOOKINGS is the agency's revenue-generating transaction, an event entity, and you need a table for that business transaction. Consider the booking transaction next.

You know that the talent agent books a certain band into a certain club on a certain date for a certain fee, starting and ending at a certain time. From that information, you can see that the attributes of the BOOKINGS entity are Band Name, Club Name, Date, Start Time, End Time, and Fee. The BOOKINGS table and its fields are shown in Figure A-7.

BOOKINGS Field Name	Data Type
Band Name	Text
Club Name	Text
Date	Date/Time
Start Time	Date/Time
End Time	Date/Time
Fee	Currency

FIGURE A-7   The BOOKINGS table and its fields—and no designation of a primary key

Some records in the BOOKINGS table are shown in Figure A-8.

Band Name	Club Name	Date	Start Time	End Time	Fee
Heartbreakers	East End	11/21/08	19:00	23:30	$800
Heartbreakers	East End	11/22/08	19:00	23:30	$750
Heartbreakers	West End	11/28/08	13:00	18:00	$500
Lightmetal	East End	11/21/08	13:00	18:00	$700
Lightmetal	West End	11/22/08	13:00	18:00	$750

FIGURE A-8 Records in the BOOKINGS table

Note that no single field is guaranteed to have unique values, because each band is likely to be booked many times and each club used many times. Further, each date and time also can appear more than once. Thus, no one field can be the primary key.

If a table does not have a single primary key field, you can make a compound primary key whose field values, when taken together, will be unique. Because a band can be in only one place at a time, one possible solution is to create a compound key consisting of the fields Band Name, Date, and Start Time. An alternative solution is to create a compound primary key consisting of the fields Club Name, Date, and Start Time.

A way to avoid having a compound key is to create a field called Booking Number. Each booking would have its own unique number, similar to an invoice number.

Here is another way to think about this event entity: Over time, a band plays in many clubs and each club hires many bands. Thus, the BAND-to-CLUB relationship is a many-to-many relationship. Such relationships signal the need for a table between the two entities in the relationship. Here you would need the BOOKINGS table, which associates the BAND and CLUB tables. An associative table is implemented by including the primary keys from the two tables that are associated. In this case, the primary keys from the BAND and CLUB tables are included as foreign keys in the BOOKINGS table.

Rule 5: Avoid data redundancy. You should not include extra (redundant) fields in a table. Doing so takes up extra disk space and leads to data entry errors because the same value must be entered in multiple tables, increasing the chance of a keystroke error. In large databases, keeping track of multiple instances of the same data is nearly impossible and contradictory data entries become a problem.

Consider this example: Why wouldn't Club Phone Number be in the BOOKINGS table as a field? After all, the agent might have to call about some last-minute change for a booking and could quickly look up the number in the BOOKINGS table. Assume the BOOKINGS table had Booking Number as the primary key and Club Phone Number as a field. Figure A-9 shows the BOOKINGS table with the additional field.

BOOKINGS	
Field Name	Data Type
Booking Number (primary key)	Text
Band Name	Text
Club Name	Text
Club Phone Number	Text
Date	Date/Time
Start Time	Date/Time
End Time	Date/Time
Fee	Currency

FIGURE A-9 The BOOKINGS table with an unnecessary field—Club Phone Number

The fields Date, Start Time, End Time, and Fee logically depend on the Booking Number primary key—they help define the booking. Band Name and Club Name are foreign keys and are needed to establish the relationship between the tables BAND, CLUB, and BOOKINGS. But what about Club Phone Number? It is not defined by the Booking Number. It is defined by Club Name—*that is, it's a function of the club, not of the booking*. Thus, the Club Phone Number field does not belong in the BOOKINGS table. It's already in the CLUB table; if the agent needs the Club Phone Number field, he or she can look it up there.

Perhaps you can see the practical data entry problem of including Club Phone Number in BOOKINGS. Suppose a club changed its contact phone number. The agent can easily change the number one time, in CLUB. But now the agent would need to remember the names of all of the other tables that had that field and change the values there too. Of course, with a small database, that might not be difficult. But in large databases, having many redundant fields in many tables makes this sort of maintenance very difficult, which means that redundant data is often incorrect.

You might object, saying, "What about all of those foreign keys? Aren't they redundant?" In a sense, they are. But they are needed to establish the relationship between one entity and another, as discussed previously.

Rule 6: *Do not include a field if it can be calculated from other fields.* A calculated field is made using the query generator. Thus, the agent's fee is not included in the BOOKINGS table, because it can be calculated by query (here, 5 percent times the booking fee).

PRACTICE DATABASE DESIGN PROBLEM

Imagine this scenario: Your town has a library that wants to keep track of its business in a database, and you have been called in to build the database. You talk to the town librarian, review the old paper-based records, and watch people use the library for a few days. You learn these things about the library:

1. Anyone who lives in the town can get a library card if he or she asks for one. The library considers each person who gets a card a "member" of the library.

2. The librarian wants to be able to contact members by telephone and by mail. She calls members when their books are overdue or when requested materials become available. She likes to mail a thank-you note to each patron on his or her anniversary of becoming a member of the library. Without a database, contacting members can be difficult to do efficiently; for example, there could be more than one member by the name of Martha Jones. Also, a parent and a child often have the same first and last name, live at the same address, and share a phone.

3. The librarian tries to keep track of each member's reading interests. When new books come in, the librarian alerts members whose interests match those books. For example, long-time member Sue Doaks is interested in reading Western novels, growing orchids, and baking bread. There must be some way to match her interests with available books. One complication is that although the librarian wants to track all of a member's reading interests, she wants to classify each book as being in just one category of interest. For example, the classic gardening book *Orchids of France* would be classified as a book about orchids or a book about France, but not both.

4. The library stocks many books. Each book has a title and any number of authors. Also, there could conceivably be more than one book in the library titled *History of the United States*. Similarly, there could be more than one author with the same name.

5. A writer could be the author of more than one book.

6. A book could be checked out repeatedly as time goes on. For example, *Orchids of France* could be checked out by one member in March, another member in July, and yet another member in September.

7. The library must be able to identify whether a book is checked out.

8. A member can check out any number of books in a visit. It's also conceivable that a member could visit the library more than once a day to check out books—some members do.

9. All books that are checked out are due back in two weeks, with no exceptions. The late fee is 50 cents per day. The librarian would like to have an automated way of generating an overdue book list each day so she can telephone the miscreants.

10. The library has a number of employees. Each employee has a job title. The librarian is paid a salary, but other employees are paid by the hour. Employees clock in and out each day. Assume all employees work only one shift per day and all are paid weekly. Pay is deposited directly into an employee's checking account—no checks are hand-delivered. The database needs to include the librarian and all other employees.

Design the library's database, following the rules set forth in this tutorial. Your instructor will specify the format of your work. Here are a few hints in the form of questions:

- A book can have more than one author. An author can write more than one book. How would you describe the relationship between books and authors?
- The library lends books for free, of course. If you were to think of checking out a book as a sale transaction for zero revenue, how would you handle the library's revenue-generating event?
- A member can borrow any number of books at a checkout. A book can be checked out more than once. How would you describe the relationship between checkouts and books?

TUTORIAL **B**

MICROSOFT ACCESS TUTORIAL

Microsoft Access is a relational database package that runs on the Microsoft Windows operating system. This tutorial was prepared using Access 2007.

Before using this tutorial, you should be familiar with the fundamentals of Microsoft Access and know how to use Windows. This tutorial teaches you some of the advanced Access skills you'll need to do database case studies. The tutorial concludes with a discussion of common Access problems and explains how to solve them.

A preliminary word of caution: always observe proper file-saving and closing procedures. Use these steps to exit Access: (1) Office Button—Close Database, then (2) X-Exit Access Button. Or you may simply click the **X-Exit Access** Button, which gets you back to Windows. Always end your work with those file-closing steps. Do not remove your disk, CD, or other portable storage device such as a USB flash drive when database forms, tables, etc., appear on the screen; you will lose your work.

To begin this tutorial, you will create a new database called **Employee**.

AT THE KEYBOARD

Open a new database. On the Getting Started with Microsoft Office Access page, under the heading New Blank Database, click Blank Database. Call the database **Employee**. Click the file folder next to the filename to browse for the desired folder. Otherwise, your file will be saved automatically in My Documents.

Your opening screen should resemble the screen shown in Figure B-1.

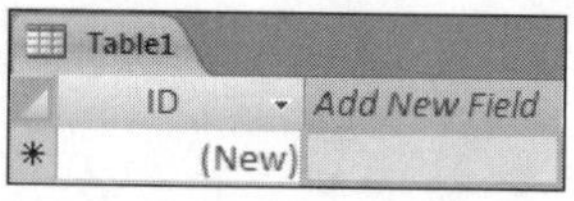

FIGURE B-1   Entering data in the Datasheet view

When you create a table, Access opens the table in the Datasheet view by default. Because you will be using the Design View to build your tables, close the new table by clicking the *x* in the upper-right corner of the table window that corresponds to Close Table I. You are now on the Home tab in the Database window of Access, as shown in Figure B-2. From this screen, you can create or change objects.

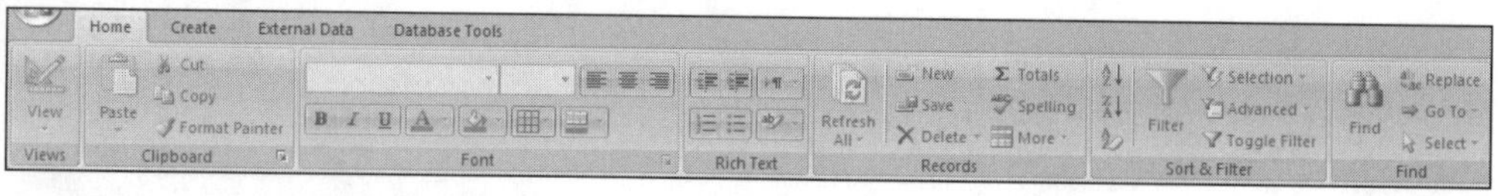

FIGURE B-2   The Database window Home tab in Access

CREATING TABLES

Your database will contain data about employees, their wage rates, and the hours they worked.

Defining Tables

In the Database window, build three new tables using the instructions that follow.

(1) Define a table called EMPLOYEE.

This table contains permanent data about employees. To create it, choose the Create tab and in the Tables group, click **Table design**. The table's fields are Last Name, First Name, SSN (Social Security Number), Street Address, City, State, Zip, Date Hired, and US Citizen. The field SSN is the primary key field. Change the length of text fields from the default 255 spaces to more appropriate lengths; for example, the field Last Name might be 30 spaces and the Zip field might be 10 spaces. Your completed definition should resemble the one shown in Figure B-3.

Field Name	Data Type	Description
Last Name	Text	
First Name	Text	
SSN	Text	
Street Address	Text	
City	Text	
State	Text	
Zip	Text	
Date Hired	Date/Time	
US Citizen	Yes/No	

FIGURE B-3 Fields in the EMPLOYEE table

When you're finished, choose Office Button—Save; then enter the name desired for the table (here, EMPLOYEE). Make sure you specify the name of the *table*, not the database itself. (Here, it is a coincidence that the EMPLOYEE table has the same name as its database file.) Close the table by clicking the x-Close corresponding to the table EMPLOYEE.

(2) Define a table called WAGE DATA.

This table contains permanent data about employees and their wage rates. The table's fields are SSN, Wage Rate, and Salaried. The field SSN is the primary key field. Use the data types shown in Figure B-4. Your definition should resemble the one shown in Figure B-4.

Field Name	Data Type	Description
SSN	Text	
Wage Rate	Currency	
Salaried	Yes/No	

FIGURE B-4 Fields in the WAGE DATA table

Use Office Button—Save to save the table definition. Name the table WAGE DATA.

(3) Define a table called HOURS WORKED.

The purpose of this table is to record the number of hours employees work each week during the year. The table's fields are SSN (text), Week # (number—long integer), and Hours (number—double). The SSN and Week # are the compound keys.

In the following example, the employee with SSN 089-65-9000 worked 40 hours in Week 1 of the year and 52 hours in Week 2.

SSN	Week	#Hours
089-65-9000	1	40
089-65-9000	2	52

Note that no single field can be the primary key field. Why? The reason is that 089-65-9000 is an entry for each week. In other words, if this employee works each week of the year, at the end of the year, 52 records will have the SSN value. Thus, SSN values will not distinguish records. In addition, no other single field can distinguish these records either, because other employees will have worked during the same week number and some employees will have worked the same number of hours. For example, 40 hours—corresponding to a full-time workweek—would be a common entry for many weeks.

However, that presents a problem because in Access, a table must have a primary key field. What is the solution? Use a compound primary key; that is, use values from more than one field to create a combined field that will distinguish records. The best compound key to use for the current example consists of the field SSN and the Week # field. Why? Because as each person works each week, the week passes. That means, for example, that there is only *one* combination of SSN 089-65-9000 and Week # 1. Because those values *can occur in only one record*, the combination distinguishes that record from all others.

How do you set a compound key? The first step is to highlight the fields in the key. Those fields must appear one after the other in the table definition screen. (Plan ahead for that format.) As an alternative, you can highlight one field, hold down the Control key, and highlight the next field.

AT THE KEYBOARD

For the HOURS WORKED table, click the first field's left prefix area (known as the row selector), hold down the button, and drag down to highlight names of all fields in the compound primary key. Your screen should resemble the one shown in Figure B-5.

Field Name	Data Type	Description
SSN	Text	
Week #	Number	
Hours	Number	

FIGURE B-5   Selecting fields for the compound primary key for the HOURS WORKED table

Now click the Key icon. Your screen should resemble the one shown in Figure B-6.

Field Name	Data Type	Description
SSN	Text	
Week #	Number	
Hours	Number	

FIGURE B-6   The compound primary key for the HOURS WORKED table

That completes the creation of the compound primary key and the table definition. Use Office Button—Save to save the table as HOURS WORKED.

Adding Records to a Table

At this point, you have set up the skeletons of three tables. The tables have no data records yet. If you were to print the tables, all you would see would be column headings (the field names). The most direct way to enter data into a table is to double-click the table's name in the Navigation Pane along the left side of the screen and type the data directly into the cells. (*Note:* Access 2007 uses a Navigation Pane to display and open the database objects. The Navigation Pane is on the left side of the Access window.)

AT THE KEYBOARD

On the Database window's Home tab, double-click the EMPLOYEE table. Your data entry screen should resemble the one shown in Figure B-7.

The EMPLOYEE table has many fields, some of which may be off the screen to the right. Scroll to see obscured fields. (Scrolling happens automatically as data is entered.) Figure B-7 shows all of the fields on the screen.

Employee

Last Name	First Name	SSN	Street Addre	City	State	Zip	Date Hired	US Citizen
*								☐

FIGURE B-7 The data entry screen for the EMPLOYEE table

Type in your data one field value at a time. Note that the first row is empty when you begin. Each time you finish a value, press Enter; the cursor will move to the next cell. After data has been entered in the last cell in a row, the cursor moves to the first cell of the next row *and* Access automatically saves the record. (Thus, there is no need to perform the Office Button—Save step after entering data into a table.)

When entering data in your table, note that dates (for example, in the Date Hired field) should be entered as follows: 6/15/07. Access automatically expands the entry to the proper format in output.

Also note that Yes/No variables are clicked (checked) for Yes; otherwise, the box is left blank for No. You can change the box from Yes to No by clicking it, as if you were using a toggle switch.

If you make errors in data entry, click the cell, backspace over the error, and type the correction.

Enter the data shown in Figure B-8 into the EMPLOYEE table.

Employee

Last Name	First Name	SSN	Street Address	City	State	Zip	Date Hired	US Citizen
Howard	Jane	114-11-2333	28 Sally Dr	Glasgow	DE	19702	8/1/2008	☑
Smith	John	123-45-6789	30 Elm St	Newark	DE	19711	6/1/1996	☑
Smith	Albert	148-90-1234	44 Duce St	Odessa	DE	19722	7/15/1987	☑
Jones	Sue	222-82-1122	18 Spruce St	Newark	DE	19716	7/15/2004	☐
Ruth	Billy	714-60-1927	1 Tater Dr	Baltimore	MD	20111	8/15/1999	☐
Add	Your	Data	Here	Elkton	MD	21921		☑
*								☐

FIGURE B-8 Data for the EMPLOYEE table

Note that the sixth record is *your* data record. Assume that you live in Elkton, Maryland, were hired on today's date (enter the date), and are a U.S. citizen. Make up a fictitious Social Security number. (For purposes of this tutorial, this sixth record, as you'll discover as you continue reading, has been created using the name of one of this text's authors and the SSN 099-11-3344.)

After adding records to the EMPLOYEE table, open the WAGE DATA table and enter the data shown in Figure B-9.

Wage Data

SSN	Wage Rate	Salaried
114-11-2333	$10.00	☐
123-45-6789		☑
148-90-1234	$12.00	☐
222-82-1122		☑
714-60-1927		☑
Your SSN	$8.00	☐
*		☐

FIGURE B-9 Data for the WAGE DATA table

In this table, you are again asked to create a new entry. For this record, enter your own SSN. Also assume that you earn $8 an hour and are not salaried. (Note that when an employee's Salaried box is not checked (i.e., Salaried = No), the implication is that the employee is paid by the hour. Because employees who are salaried do not get paid by the hour, their hourly rate is shown as 0.00.)

Once you have finished creating the WAGE DATA table, open the HOURS WORKED table and enter the data shown in Figure B-10.

Notice that salaried employees are always given 40 hours. Nonsalaried employees (including you) might work any number of hours. For your record, enter your fictitious SSN, 60 hours worked for Week 1, and 55 hours worked for Week 2.

Hours Worked

SSN	Week #	Hours
114-11-2333	1	40
114-11-2333	2	50
123-45-6789	1	40
123-45-6789	2	40
148-90-1234	1	38
148-90-1234	2	40
222-82-1122	1	40
222-82-1122	2	40
714-60-1927	1	40
714-60-1927	2	40
Your SSN	1	60
Your SSN	2	55

FIGURE B-10 Data for the HOURS WORKED table

CREATING QUERIES

Because you know how to create basic queries, this section teaches you the kinds of advanced queries you will create in the Case Studies.

Using Calculated Fields in Queries

A **calculated field** is an output field made up of *other* field values. A calculated field is *not* a field in a table; it is created in the query generator. The calculated field does not become part of the table—it is just part of query output. The best way to understand this process is to work through an example.

AT THE KEYBOARD

Suppose you want to see the SSNs and wage rates of hourly workers and you want to see what the wage rates would be if all employees were given a 10 percent raise. To view that information, show the SSN, the current wage rate, and the higher rate (which should be titled New Rate in the output). Figure B-11 shows how to set up the query.

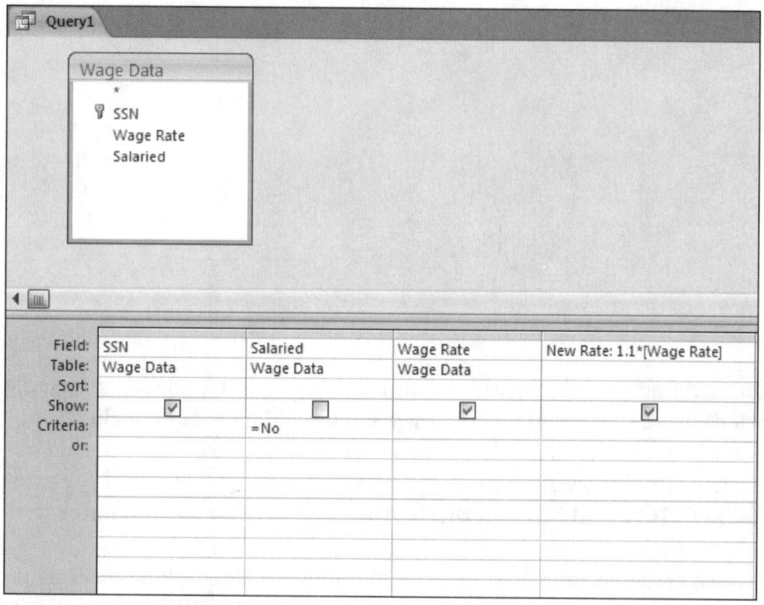

FIGURE B-11 Query setup for the calculated field

To set up this query, you need to select hourly workers by using the Salaried field with the Criteria = No. Note in Figure B-11 that the Show box for that field is not checked, so the Salaried field values will not show in the query output.

Note the expression for the calculated field, which you can see in the rightmost field cell:

New Rate: 1.1 * [Wage Rate]

The term *New Rate:* merely specifies the desired output heading. (Don't forget the colon.) The 1.1 * [Wage Rate] multiplies the old wage rate by 110 percent, which results in the 10 percent raise.

In the expression, the field name Wage Rate must be enclosed in square brackets. That is a rule. *Any time an Access expression refers to a field name, the expression must be enclosed in square brackets.*

If you run this query, your output should resemble that shown in Figure B-12.

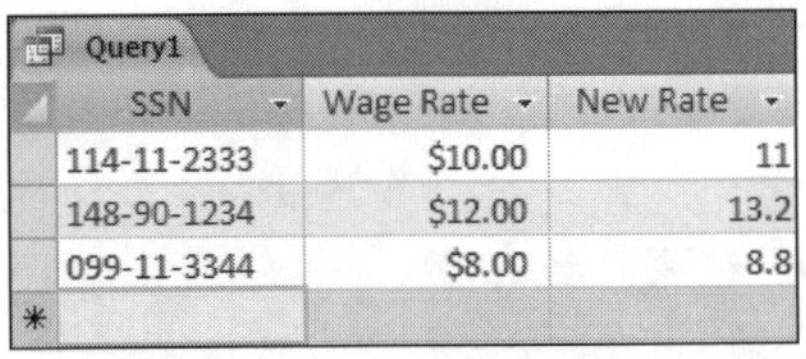

SSN	Wage Rate	New Rate
114-11-2333	$10.00	11
148-90-1234	$12.00	13.2
099-11-3344	$8.00	8.8

FIGURE B-12   Output for a query with calculated field

Notice that the calculated field output is not shown in Currency format; it's shown as a Double—a number with digits after the decimal point. To convert the output to Currency format, you should select the output column by clicking the line above the calculated field expression, thus activating the column, which subsequently darkens. Your data entry screen should resemble the one shown in Figure B-13.

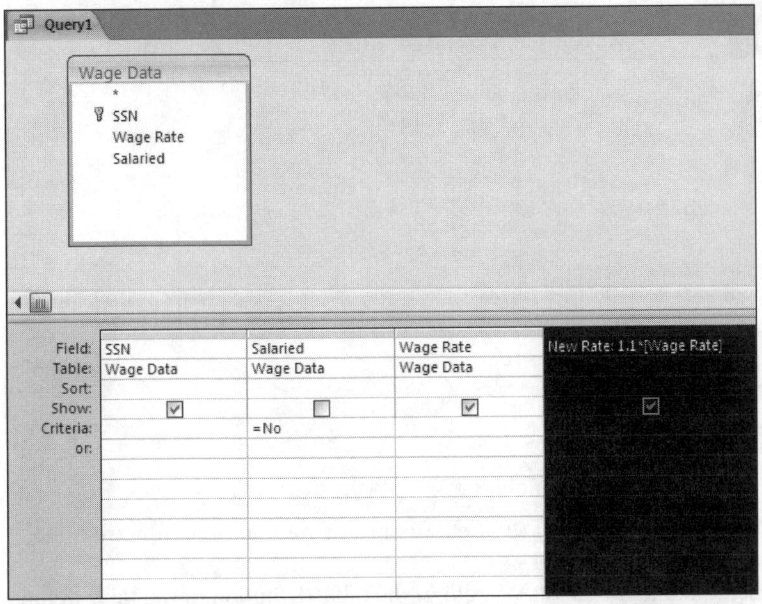

FIGURE B-13   Activating a calculated field in query design

Then on the Design tab header, click **Property Sheet** in the Show/Hide group. A Field Properties window appears, as shown on the right in Figure B-14.

Click **Format** and choose Currency, as shown in Figure B-15. Then click the upper-right X to close the window.

Now when you run the query, the output should resemble that shown in Figure B-16.

Next, you'll look at how to avoid errors when making calculated fields.

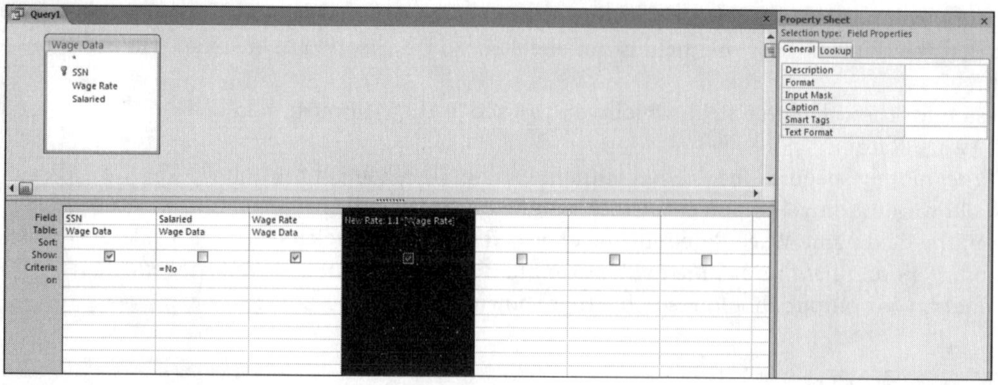

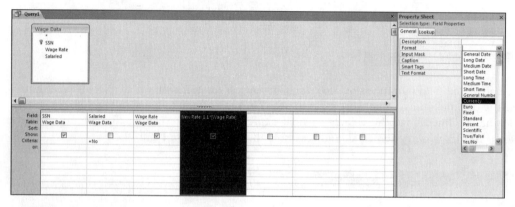

FIGURE B-14   Field Properties of a calculated field

FIGURE B-15   Currency format of a calculated field

SSN	Wage Rate	New Rate
114-11-2333	$10.00	$11.00
148-90-1234	$12.00	$13.20
099-11-3344	$8.00	$8.80

FIGURE B-16   Query output with formatted calculated field

Avoiding Errors When Making Calculated Fields

Follow these guidelines to avoid making errors in calculated fields:

- Don't put the expression in the *Criteria* cell as if the field definition were a filter. You are making a field, so put the expression in the *Field* cell.
- Spell, capitalize, and space a field's name *exactly* as you did in the table definition. If the table definition differs from what you type, Access thinks you're defining a new field by that name. Access then prompts you to enter values for the new field, which it calls a Parameter Query field. That is easy to debug because of the tag Parameter Query. If Access asks you to enter values for a Parameter, you almost certainly misspelled a field name in an expression in a calculated field or criterion.

  Example: Here are some errors you might make for Wage Rate:

  Misspelling: (Wag Rate)
  Case change: (wage Rate / WAGE RATE)
  Spacing change: (WageRate / Wage Rate)

- Don't use parentheses or curly braces instead of the square brackets. Also, don't put parentheses inside square brackets. You *are*, however, allowed to use parentheses outside the square brackets, in the normal algebraic manner.

Example: Suppose you want to multiply Hours times Wage Rate to get a field called Wages Owed. This is the correct expression:

    Wages Owed: [Wage Rate] * [Hours]

The following also would be correct:

    Wages Owed: ([Wage Rate] * [Hours])

But it would *not* be correct to omit the inside brackets, which is a common error:

    Wages Owed: [Wage Rate * Hours]

"Relating" Two (or More) Tables by the Join Operation

Often the data you need for a query is in more than one table. To complete the query, you must join the tables by linking the common fields, which is known as a *join*. One rule of thumb is that joins are made on fields that have common *values*, and those fields often can be key fields. The names of the join fields are irrelevant; also, the names of the tables (or fields) to be joined may be the same, but that is not a requirement for an effective join.

Make a join by bringing in (Adding) the tables needed. Next, decide which fields you will join. Then click one field name and hold down the left mouse button while you drag the cursor over to the other field's name in its window. Release the button. Access puts in a line, signifying the join. (*Note:* If a relationship between two tables has been formed elsewhere, Access will put in the line automatically, in which case you do not have to perform the click-and-drag operation. Often Access puts in join lines without the user forming relationships.)

You can join more than two tables. The common fields *need not* be the same in all tables; that is, you can daisy-chain them together.

A common join error is to add a table to the query and then fail to link it to another table. In that case, you will have a table floating in the top part of the QBE (query by example) screen. When you run the query, your output will show the same records over and over. That error is unmistakable because there is *so much* redundant output. The rules are (1) add only the tables you need and (2) link all tables.

Next, you'll work through an example of a query needing a join.

AT THE KEYBOARD

Suppose you want to see the last names, SSNs, wage rates, salary status, and citizenship only for U.S. citizens and hourly workers. Because the data is spread across two tables, EMPLOYEE and WAGE DATA, you should add both tables and pull down the five fields you need. Then you should add the Criteria. Set up your work to resemble that shown in Figure B-17. Make sure the tables are joined on the common field, SSN.

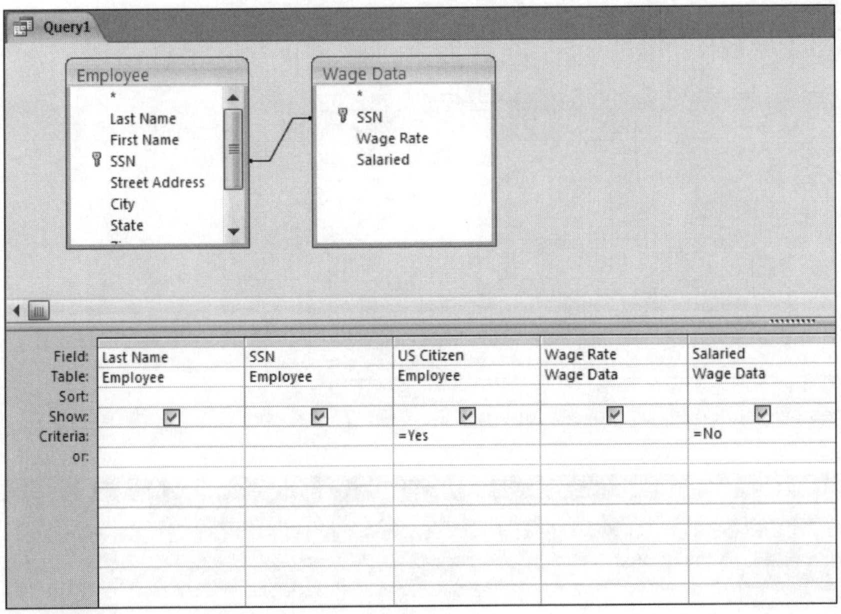

FIGURE B-17   A query based on two joined tables

Here is a quick review of the Criteria you will need to set up this join: If you want data for employees who are U.S. citizens *and* who are hourly workers, the Criteria expressions go in the *same* Criteria row. If you want data for employees who are U.S. citizens *or* who are hourly workers, one of the expressions goes in the second Criteria row (the one that has the or: notation).

Now run the query. The output should resemble that shown in Figure B-18, with the exception of the name Brady.

Last Name	SSN	US Citizen	Wage Rate	Salaried
Howard	114-11-2333	☑	$10.00	☐
Smith	148-90-1234	☑	$12.00	☐
Brady	099-11-3344	☑	$8.00	☐
*		☐		☐

FIGURE B-18   Output of a query based on two joined tables

As there is no need to print the query output or save it, you should go back to the Design View and close the query. Another practice query follows.

AT THE KEYBOARD

Suppose you want to see the wages owed to hourly employees for Week 2. To do that, you should show the last name, the SSN, the salaried status, the week #, and the wages owed. Wages will have to be a calculated field ([Wage Rate] * [Hours]). The criteria are No for Salaried and 2 for the Week #. (*Note:* This means another "And" query is required.) Your query should be set up like the query displayed in Figure B-19.

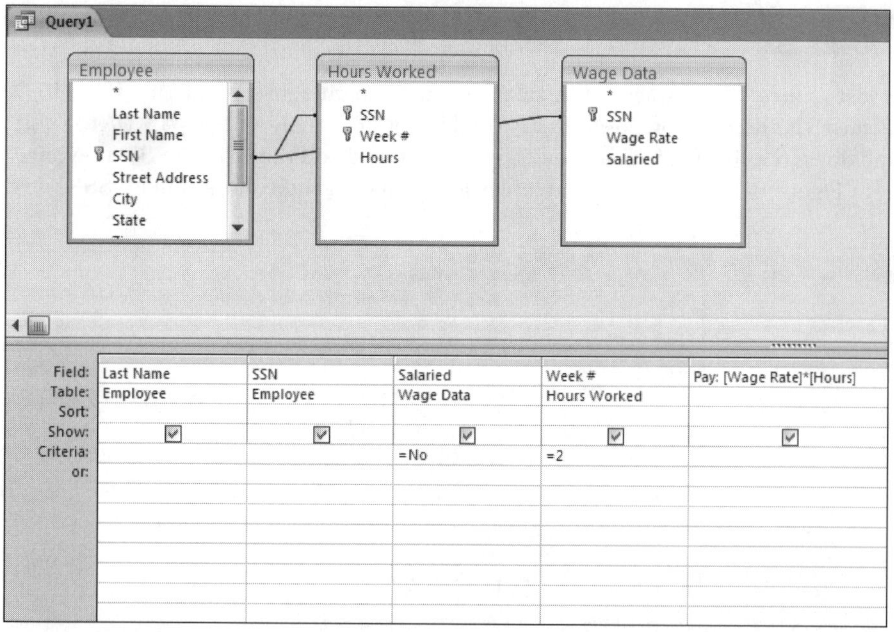

Field:	Last Name	SSN	Salaried	Week #	Pay: [Wage Rate]*[Hours]
Table:	Employee	Employee	Wage Data	Hours Worked	
Sort:					
Show:	☑	☑	☑	☑	☑
Criteria:			=No	=2	
or:					

FIGURE B-19   Query setup for wages owed to hourly employees for Week 2

NOTE

Note that in the query in Figure B-19, the calculated field column was widened so you can see the whole expression. To widen a column, click the column boundary line and drag to the right.

Run the query. The output should be similar to that shown in Figure B-20 (if you formatted your calculated field to Currency).

Last Name	SSN	Salaried	Week #	Pay
Howard	114-11-2333	☐	2	$500.00
Smith	148-90-1234	☐	2	$480.00
Brady	099-11-3344	☐	2	$440.00
*		☐		

FIGURE B-20   Query output for wages owed to hourly employees for Week 2

Notice that it was not necessary to pull down the Wage Rate and Hours fields to make the query work. As there is no need to print the query output or save it, you should go back to the Design View and close the query.

Summarizing Data from Multiple Records (Totals Queries)

You may want data that summarizes values from a field for several records (or possibly all records) in a table. For example, you might want to know the average hours that all employees worked in a week or perhaps the total (sum) of all of the hours worked. Furthermore, you might want data grouped (stratified) in some way. For example, you might want to know the average hours worked, grouped by all U.S. citizens versus all non-U.S. citizens. Access calls that kind of query a **Totals query**. Those operations include the following:

Sum	The total of a given field's values
Count	A count of the number of instances in a field—that is, the number of records. (In the current example, to get the number of employees, you'd count the number of SSNs.)
Average	The average of a given field's values
Min	The minimum of a given field's values
Var	The variance of a given field's values
StDev	The standard deviation of a given field's values
Where	The field has criteria for the query output

▣ AT THE KEYBOARD

Suppose you want to know how many employees are represented in the example database. The first step is to bring the EMPLOYEE table into the QBE screen. Do that now. Because you will need to count the number of SSNs, which is a Totals query operation, you must bring down the SSN field.

To tell Access you want a Totals query, click the little Totals icon in the Design tab in the Show/Hide group.

That opens a new row in the lower part of the QBE screen, called the Total row. At this point, the screen resembles that shown in Figure B-21.

Note that the Total cell contains the words *Group By*. Until you specify a statistical operation, Access assumes a field will be used for grouping (stratifying) data.

To count the number of SSNs, click next to Group By, which reveals a little arrow. Click the arrow to reveal a drop-down menu, as shown in Figure B-22.

Select the Count operator. (With this menu, you may need to scroll to see the operator you want.) Your screen should now resemble the one shown in Figure B-23.

Run the query. Your output should resemble that shown in Figure B-24.

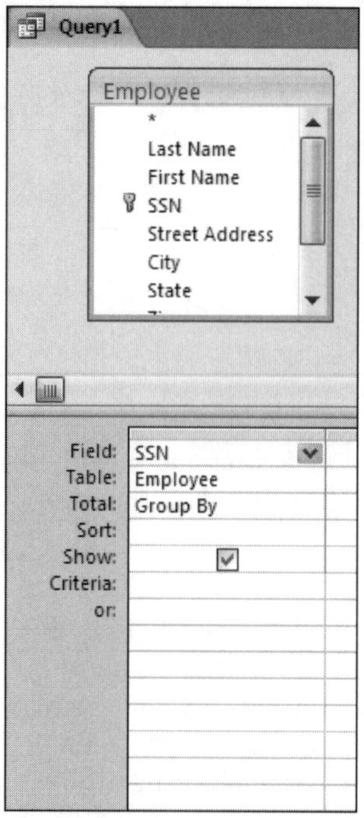

FIGURE B-21   Totals query setup

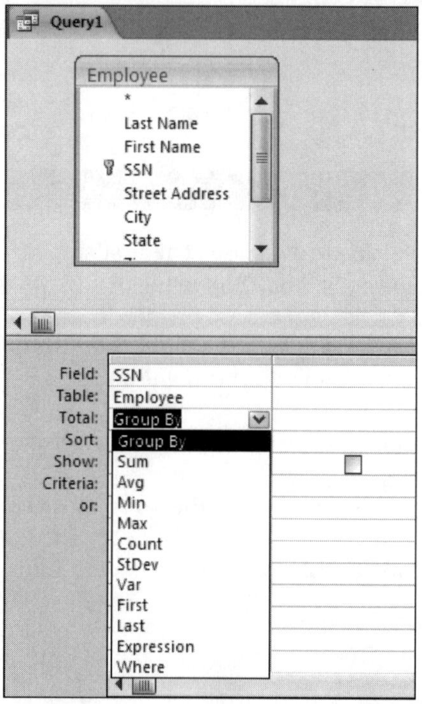

FIGURE B-22   Choices for statistical operation in a Totals query

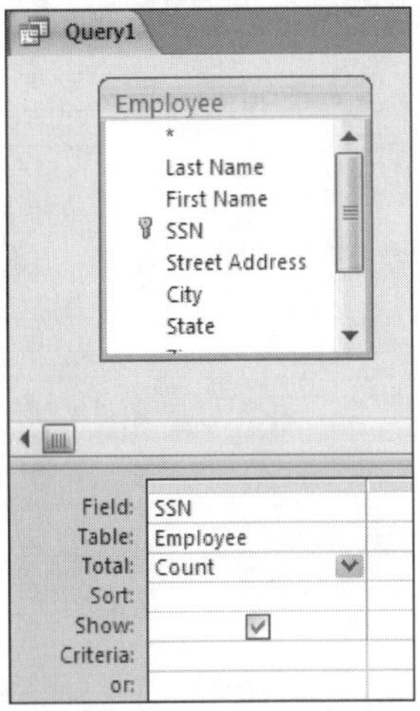

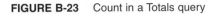

FIGURE B-23   Count in a Totals query

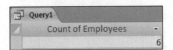

FIGURE B-24   Output of Count in a Totals query

Notice that Access made a pseudo-heading "CountOfSSN." To do that, Access spliced together the statistical operation (Count), the word *Of*, and the name of the field (SSN). What if you wanted a phrase such as *Count of Employees* as a heading? In the Design View, you'd change the query to resemble the one shown in Figure B-25.

FIGURE B-25   Heading change in a Totals query

Now when you run the query, the output should resemble that shown in Figure B-26.

FIGURE B-26   Output of heading change in a Totals query

As there is no need to print the query output or save it, you should go back to the Design View and close the query.

Here is another example of a Totals query. Suppose you want to know the average wage rate of employees, grouped by whether the employees are salaried.

Figure B-27 shows how your query should be set up.

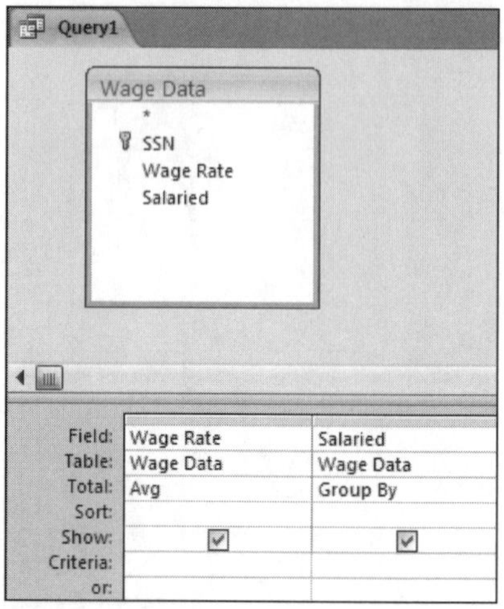

FIGURE B-27   Query setup for average wage rate of employees

When you run the query, your output should resemble that shown in Figure B-28.

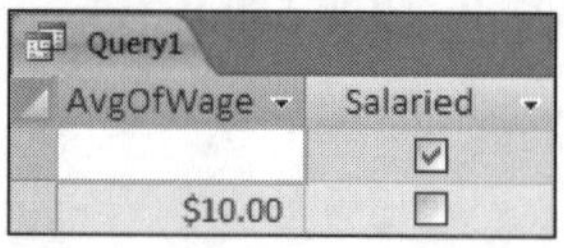

FIGURE B-28   Output of query for average wage rate of employees

Recall the convention that salaried workers are assigned zero dollars an hour. Suppose you want to eliminate the output line for zero dollars an hour because only hourly rate workers matter for that query. The query setup is shown in Figure B-29.

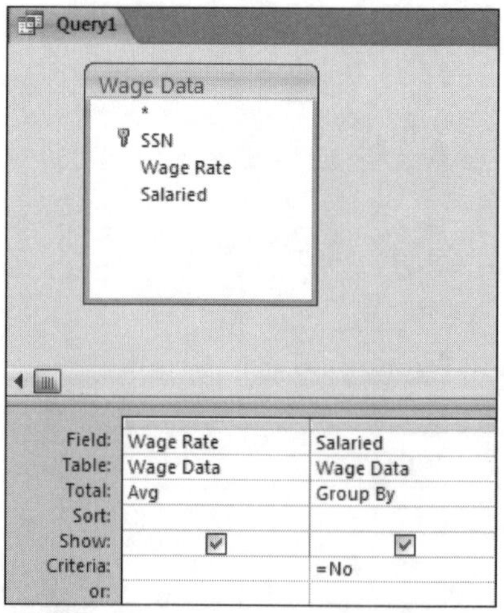

FIGURE B-29   Query setup for nonsalaried workers only

Now when you run the query, you'll get output for nonsalaried employees only, as shown in Figure B-30.

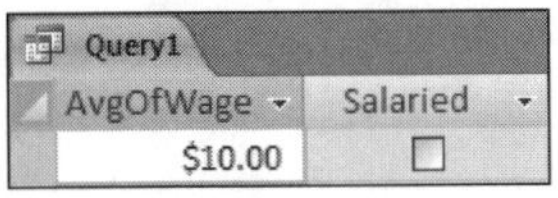

FIGURE B-30   Query output for nonsalaried workers only

Thus, it's possible to use Criteria in a Totals query, just as you would with a "regular" query. As there is no need to print the query output or save it, you should go back to the Design View and close the query.

AT THE KEYBOARD

Assume you want to see two pieces of information for hourly workers: (1) the average wage rate—call it Average Rate in the output and (2) 110 percent of the average rate—call it the Increased Rate. To do that, you can make a calculated field in a Totals query.

You already know how to do certain things for this query. The revised heading for the average rate will be Average Rate (type *Average Rate: Wage Rate* in the Field cell). Note that you want the average of this field. Also, the grouping would be by the Salaried field. (To get hourly workers only, enter *Criteria: No.*)

The most difficult part of this query is to construct the expression for the calculated field. Conceptually, it is as follows:

Increased Rate: 1.1 * [The current average, however that is denoted]

The question is how to represent [The current average]. You cannot use Wage Rate, because that heading denotes the wages before they are averaged. Surprisingly, you can use the new heading (Average Rate) to denote the averaged amount as follows:

Increased Rate: 1.1 * [Average Rate]

Although it may seem counterintuitive, *you can treat* Average Rate *as if it were an actual field name*. Note, however, that if you use a calculated field such as Average Rate in another calculated field, as shown in Figure B-31, you must show that original calculated field in the query output. If you don't, the query will ask you to *enter parameter value*, which is incorrect. Use the setup shown in Figure B-31.

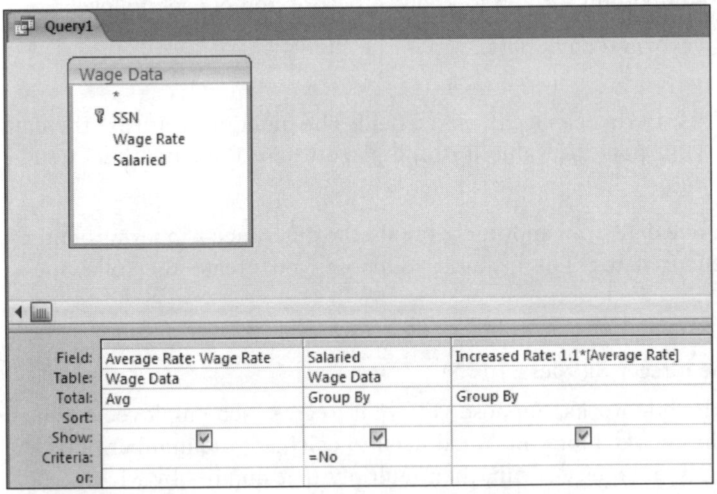

FIGURE B-31   Using a calculated field in another calculated field

NOTE

If you ran the query now shown in Figure B-31, you'd get some sort of error message because there is no *statistical* operator applied to the calculated field's Total cell; instead, the words *Group By* appear there. To correct that, you must change the Group By operator to Expression. You may have to scroll down to get to Expression in the list.

Figure B-32 shows how the screen looks before the query is run.
Figure B-33 shows the output of the query.

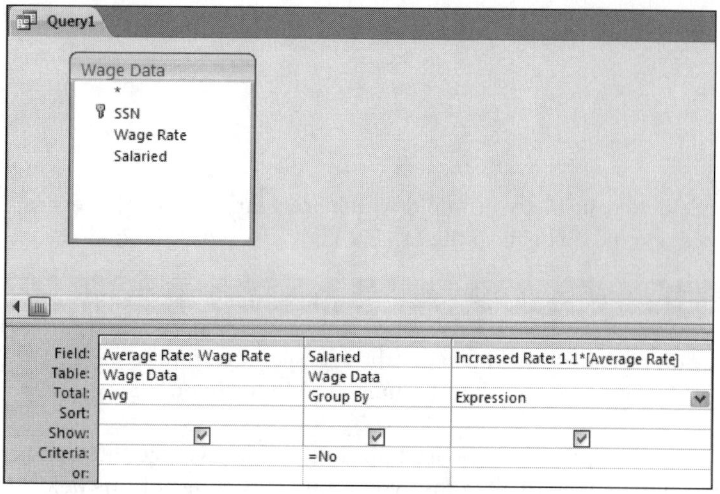

FIGURE B-32   An Expression in a Totals query

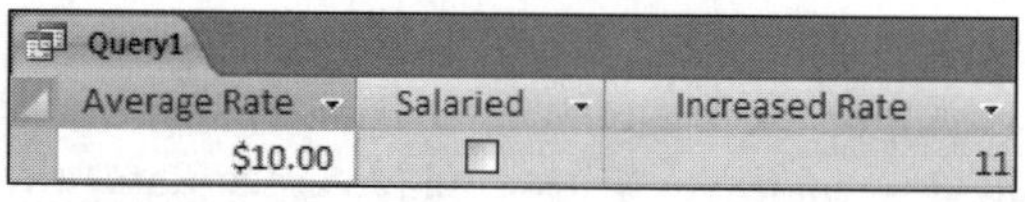

FIGURE B-33   Output of an Expression in a Totals query

As there is no need to print the query output or save it, you should go back to the Design View and close the query.

Using the Date() Function in Queries

Access has two date function features that are important for you to know. A description of each follows.

1. The following built-in function gives you today's date:

    Date()

    You can use that function in query criteria or in a calculated field. The function "returns" the day on which the query is run; that is, it puts the value into the place where the date() function appears in an expression.

2. *Date arithmetic* lets you subtract one date from another to obtain the difference—in terms of number of days—between two calendar dates. For example, suppose you create the following expression:

    10/9/2007 – 10/4/2007

    Access would evaluate that as the integer 5 (9 less 4 is 5).

Here is another example of how date arithmetic works. Suppose you want to give each employee a bonus equaling a dollar for each day the employee has worked for you. You'd need to calculate the number of days between the employee's date of hire and the day the query is run, then multiply that number by $1.

You would find the number of elapsed days by using the following equation:

Date() – [Date Hired]

Also suppose that for each employee, you want to see the last name, SSN, and bonus amount. You'd set up the query as shown in Figure B-34.

Assume you had set the format of the Bonus field to Currency. The output will be similar to that shown in Figure B-35. (Your Bonus data will be different because you are working on a date that is different from the date this tutorial was written.)

Using Time Arithmetic in Queries

Access also allows you to subtract the values of time fields to get an elapsed time. Assume your database has a JOB ASSIGNMENTS table that shows the times that nonsalaried employees were at work during a day. The definition is shown in Figure B-36.

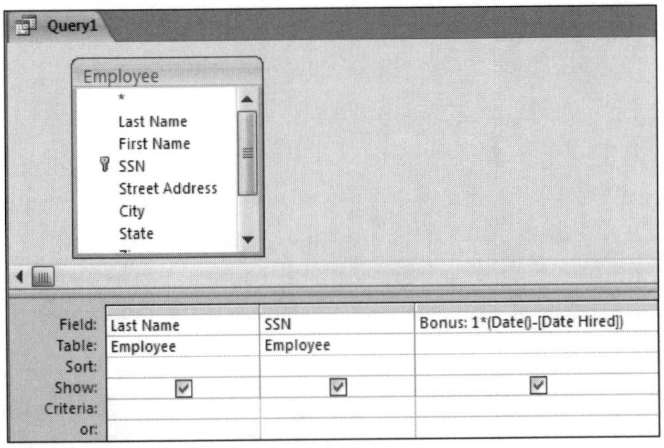

FIGURE B-34   Date arithmetic in a query

Field:	Last Name	SSN	Bonus: 1*(Date()-[Date Hired])
Table:	Employee	Employee	
Sort:			
Show:	☑	☑	☑
Criteria:			
or:			

FIGURE B-35   Output of query with date arithmetic

Last Name	SSN	Bonus
Brady	099-11-3344	$0.00
Howard	114-11-2333	$179.00
Smith	123-45-6789	$3,892.00
Smith	148-90-1234	$7,136.00
Jones	222-82-1122	$926.00
Ruth	714-60-1927	$2,722.00

FIGURE B-36   Date/Time data definition in the JOB ASSIGNMENTS table

Field Name	Data Type
SSN	Text
ClockIn	Date/Time
ClockOut	Date/Time
DateWorked	Date/Time

Assume the Date field is formatted for Long Date and the ClockIn and ClockOut fields are formatted for Medium Time. Also assume that for a particular day, nonsalaried workers were scheduled as shown in Figure B-37.

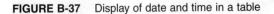

SSN	ClockIn	ClockOut	DateWorked	Add New Field
099-11-3344	8:30 AM	4:30 PM	Tuesday, September 30, 2008	
114-11-2333	9:00 AM	3:00 PM	Tuesday, September 30, 2008	
148-90-1234	7:00 AM	5:00 PM	Tuesday, September 30, 2008	

FIGURE B-37   Display of date and time in a table

You want a query that will show the elapsed time that your employees were on the premises for the day. When you add the tables, your screen may show the links differently. Click and drag the JOB ASSIGN-MENTS, EMPLOYEE, and WAGE DATA table icons to look like those in Figure B-38.

Figure B-39 shows the output, which looks correct. For example, employee 099-11-3344 was at work from 8:30 a.m. to 4:30 p.m., which is eight hours. But how does the odd expression that follows yield the correct answers?

([ClockOut] – [ClockIn]) * 24

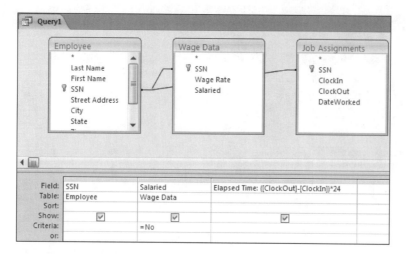

FIGURE B-38 Query setup for time arithmetic

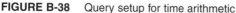

FIGURE B-39 Query output for time arithmetic

Why wouldn't the following expression work?

[ClockOut] – [ClockIn]

Here is the answer: In Access, subtracting one time from the other yields the *decimal* portion of a 24-hour day. Returning to the example, you can see that employee 099-11-3344 worked eight hours, which is one-third of a day, so the time arithmetic function yields .3333. That is why you must multiply by 24—to convert from decimals to an hourly basis. Hence, for employee 099-11-3344, the expression performs the following calculation: 1/3 × 24 = 8.

Note that parentheses are needed to force Access to do the subtraction *first*, before the multiplication. Without parentheses, multiplication takes precedence over subtraction. For example, consider the following expression:

[ClockOut] – [ClockIn] * 24

In that case, ClockIn would be multiplied by 24, the resulting value would be subtracted from ClockOut, and the output would be a nonsensical decimal number.

Deleting and Updating Queries

The queries presented in this tutorial thus far have been Select queries. They select certain data from specific tables based on a given criterion. You also can create queries to update the original data in a database. Businesses do that often, and they do it in real time. For example, when you order an item from a Web site, the company's database is updated to reflect the purchase of the item through the deletion of that item from the company's inventory.

Consider an example. Suppose you want to give all of the nonsalaried workers a $0.50 per hour pay raise. With the three nonsalaried workers you have now, it would be easy to go into the table and simply change the Wage Rate data. But assume you have 3,000 nonsalaried employees. Now it would be much faster—not to mention more accurate—to change the Wage Rate data for each of the 3,000 nonsalaried employees by using an Update query that adds $0.50 to each employee's wage rate.

Now you will change each of the nonsalaried employees' pay via an Update query. Figure B-40 shows how to set up the query.

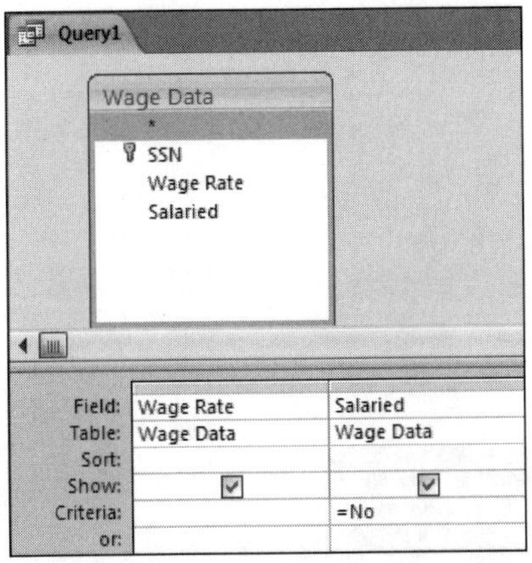

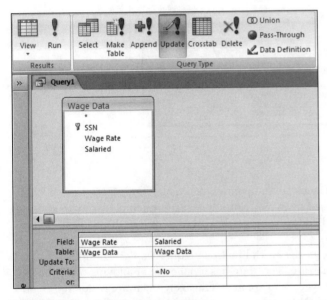

FIGURE B-40   Query setup for an Update query

So far, this query is just a Select query. Click the Update button in the Query Type group, as shown in Figure B-41.

FIGURE B-41   Selecting a query type

Notice that you now have another line on the QBE grid called *Update to:*. That is where you specify the change or update to the data. Notice that you will update only the nonsalaried workers by using a filter under the Salaried field. Update the Wage Rate data to Wage Rate plus $0.50, as shown in Figure B-42. (Note that the update involves the use of brackets [ ], as in a calculated field.)

Now run the query by clicking the Run button in the Results group. If you are unable to run the query because it's blocked by Disabled Mode, you need to choose Database Tools tab, Message Bar in the Show/Hide group. Click the Options button and choose "enable this content", then **OK**. When you successfully run the query, you will get a warning message as shown in Figure B-43.

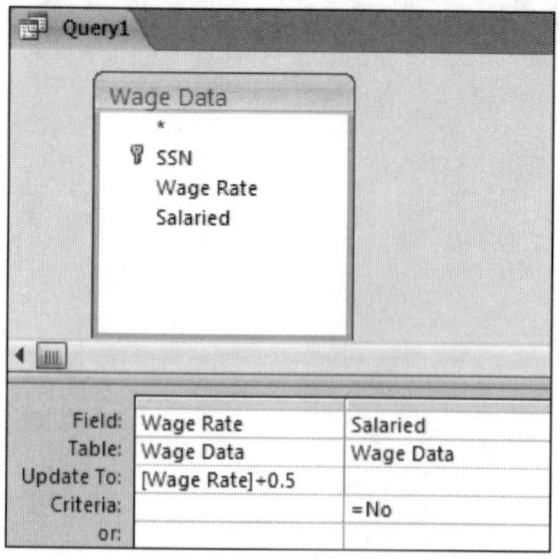

FIGURE B-42 Updating the wage rate for nonsalaried workers

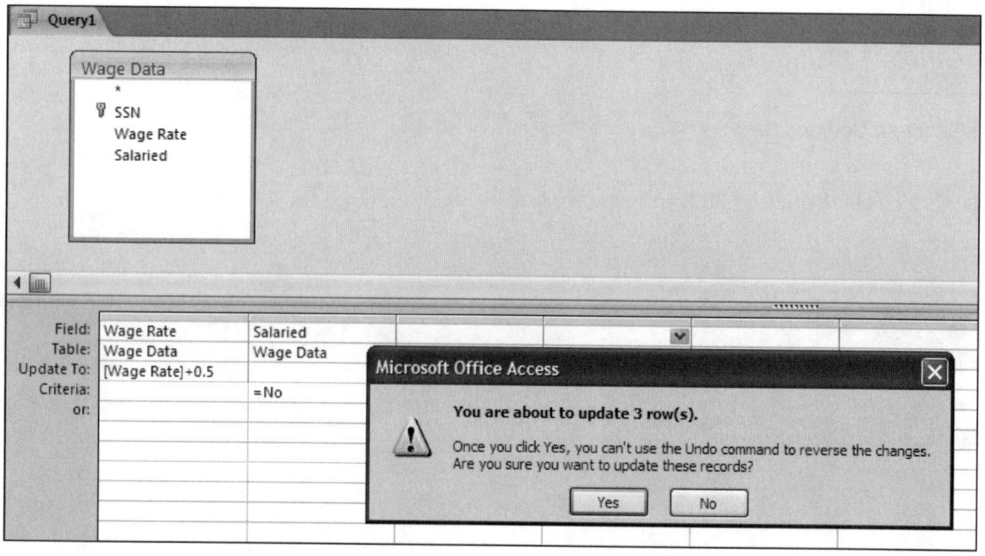

FIGURE B-43 Update Query warning

Once you click **Yes**, the records will be updated. Check those updated records now by viewing the WAGE DATA table. Each salaried wage rate should be increased by $0.50. Note that in this example, you are simply adding $0.50 to each salaried wage rate. You could add or subtract data from another table as well. If you do, remember to put the field name in square brackets.

Another kind of query is the Delete query. Delete queries work like Update queries. For example, assume your company has been taken over by the state of Delaware. The state has a policy of employing only Delaware residents. Thus, you must delete (or fire) all employees who are not exclusively Delaware residents. To do that, you need to create a Select query. Using the EMPLOYEE table, you would choose the Delete icon from the Query Type group, then bring down the State field and filter only those records not in Delaware (DE). Do not perform the operation, but note that if you did, the setup would look like the one in Figure B-44.

Using Parameter Queries

Another kind of query, which is actually a type of Select query, is a **Parameter query**. Here is an example: Suppose your company has 5,000 employees and you want to query the database to find the same kind of information again and again, but about different employees each time. For example, you might want to query the database to find out how many hours a particular employee has worked. To do that, you could run a query that you created and stored previously, but run it only for a particular employee.

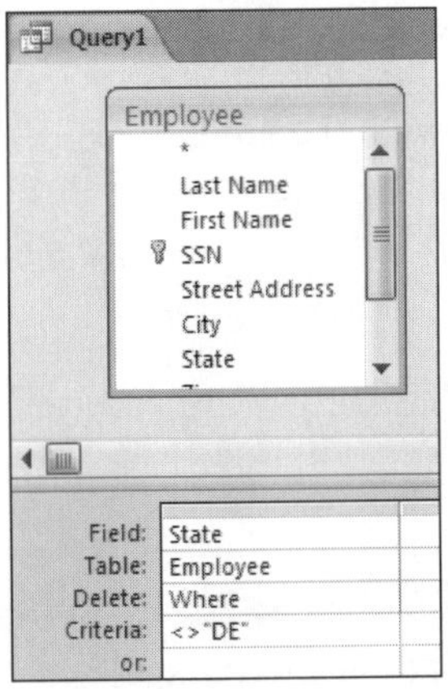

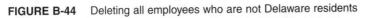

FIGURE B-44 Deleting all employees who are not Delaware residents

AT THE KEYBOARD

Create a Select query with the format shown in Figure B-45.

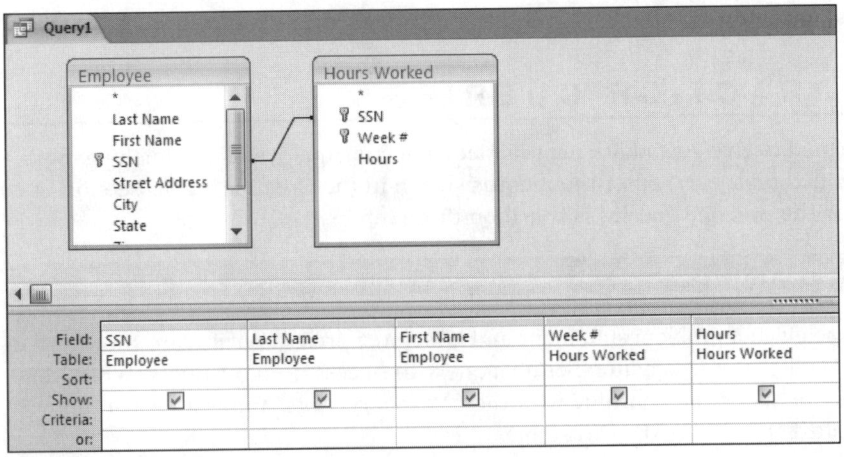

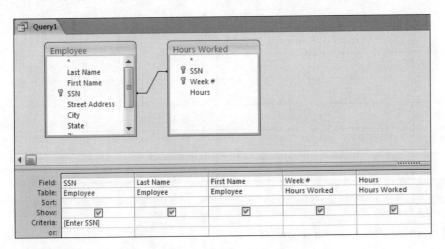

FIGURE B-45 Design of a Parameter query begins as a Select query

In the Criteria line of the QBE grid for the field SSN, type what is shown in Figure B-46.

FIGURE B-46 Design of a Parameter query, continued

Note that the Criteria line involves the use of square brackets, as you would expect to see in a calculated field.

Now run the query. You will be prompted for the specific employee's SSN, as shown in Figure B-47.

FIGURE B-47   Enter Parameter Value dialog box

Type in your own (fictitious) SSN. Your query output should resemble that shown in Figure B-48.

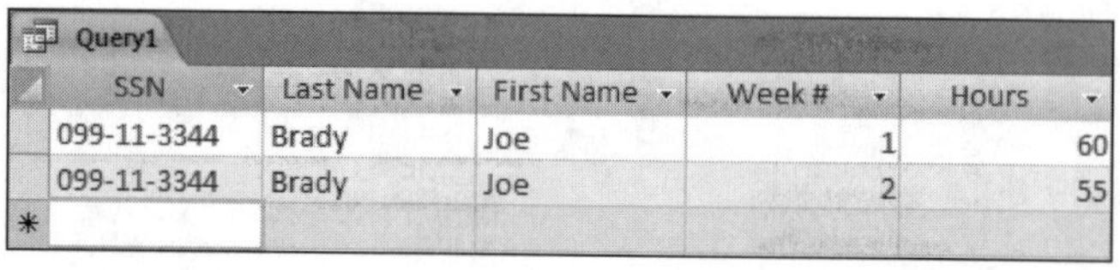

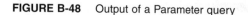

SSN	Last Name	First Name	Week #	Hours
099-11-3344	Brady	Joe	1	60
099-11-3344	Brady	Joe	2	55
*				

FIGURE B-48   Output of a Parameter query

MAKING SEVEN PRACTICE QUERIES

This portion of the tutorial is designed to give you additional practice in making queries. Before making these queries, you must create the specified tables and enter the records shown in the Creating Tables section of this tutorial. The output shown for the practice queries is based on those inputs.

AT THE KEYBOARD

For each query that follows, you are given a problem statement and a "scratch area." You also are shown what the query output should look like. Follow this procedure: Set up a query in Access. Run the query. When you are satisfied with the results, save the query and continue with the next query. *Note:* You will be working with the EMPLOYEE, HOURS WORKED, and WAGE DATA tables.

1. Create a query that shows the SSN, last name, state, and date hired for those employees who are currently living in Delaware *and* were hired after 12/31/99. Sort (ascending) by SSN. (A recap of sorting procedures: Click the Sort cell of the field. Choose Ascending or Descending.) Before creating your query, use the table shown in Figure B-49 to work out your QBE grid on paper.

Field					
Table					
Sort					
Show					
Criteria					
Or:					

FIGURE B-49   QBE grid template

Your output should resemble that shown in Figure B-50.

SSN	Last Name	State	Date Hired
114-11-2333	Howard	DE	8/1/2008
222-82-1122	Jones	DE	7/15/2004
*			

FIGURE B-50 Number 1 query output

2. Create a query that shows the last name, first name, date hired, and state for those employees who are currently living in Delaware *or* were hired after 12/31/99. The primary sort (ascending) is on last name, and the secondary sort (ascending) is on first name. (Sorting recap: The Primary Sort field must be to the left of the Secondary Sort field in the query setup.) Before creating your query, use the table shown in Figure B-51 to work out your QBE grid on paper.

Field					
Table					
Sort					
Show					
Criteria					
Or:					

FIGURE B-51 QBE grid template

If your name was Joe Brady, your output would look like that shown in Figure B-52.

Last Name	First Name	Date Hired	State
Brady	Joe	9/14/2007	MD
Howard	Jane	8/1/2008	DE
Jones	Sue	7/15/2004	DE
Smith	Albert	7/15/1987	DE
Smith	John	6/1/1996	DE
*			

FIGURE B-52 Number 2 query output

3. Create a query that shows the sum of hours that U.S. citizens worked and the same sum for non-U.S. citizens (that is, group on citizenship). The heading for total hours worked should be Total Hours Worked. Before creating your query, use the table shown in Figure B-53 to work out your QBE grid on paper.

Field					
Table					
Total					
Sort					
Show					
Criteria					
Or:					

FIGURE B-53   QBE grid template

Your output should resemble that shown in Figure B-54.

Total Hours Worked	US Citizen
363	☑
160	☐

FIGURE B-54   Number 3 query output

4. Create a query that shows the wages owed to hourly workers for Week 1. The heading for the wages owed should be Total Owed. The output headings should be Last Name, SSN, Week #, and Total Owed. Before creating your query, use the table shown in Figure B-55 to work out your QBE grid on paper.

Field					
Table					
Sort					
Show					
Criteria					
Or:					

FIGURE B-55   QBE grid template

If your name was Joe Brady, your output would look like that shown in Figure B-56.

Last Name	SSN	Week #	Total Owed
Howard	114-11-2333	1	$420.00
Smith	148-90-1234	1	$475.00
Brady	099-11-3344	1	$510.00

FIGURE B-56   Number 4 query output

5. Create a query that shows the last name, SSN, hours worked, and overtime amount owed for hourly employees who earned overtime during Week 2. Overtime is paid at 1.5 times the normal hourly rate for all of the hours worked over 40. Note that the amount shown in the query should be just the overtime portion of the wages paid. Also, this is not a Totals query—amounts should be

shown for individual workers. Before creating your query, use the table shown in Figure B-57 to work out your QBE grid on paper.

Field					
Table					
Sort					
Show					
Criteria					
Or:					

FIGURE B-57   QBE grid template

If your name was Joe Brady, your output would look like that shown in Figure B-58.

Last Name	SSN	Hours	OT Pay
Howard	114-11-2333	50	$157.50
Brady	099-11-3344	55	$191.25

FIGURE B-58   Number 5 query output

6. Create a Parameter query that shows the hours employees have worked. Have the Parameter query prompt for the week number. The output headings should be Last Name, First Name, Week #, and Hours. Do this only for the nonsalaried workers. Before creating your query, use the table shown in Figure B-59 to work out your QBE grid on paper.

Field					
Table					
Sort					
Show					
Criteria					
Or:					

FIGURE B-59   QBE grid template

Run the query using 2 when prompted for the Week #. Your output should look like that shown in Figure B-60.

Last Name	First Name	Week #	Hours
Howard	Jane	2	50
Smith	Albert	2	40
Brady	Joe	2	55

FIGURE B-60   Number 6 query output

7. Create an update query that gives certain workers a merit raise. First, you must create an additional table as shown in Figure B-61.

Merit Raises		
SSN	Merit Raise	Add New Field
114-11-2333	$0.25	
148-90-1234	$0.15	
*		

FIGURE B-61   MERIT RAISES table

Now make a query that adds the Merit Raise to the current Wage Rate for those employees who will receive a raise. When you run the query, you should be prompted with *You are about to update two rows.* Check the original WAGE DATA table to confirm the update. Before creating your query, use the table shown in Figure B-62 to work out your QBE grid on paper.

Field					
Table					
Update to					
Criteria					
Or:					

FIGURE B-62   QBE grid template

CREATING REPORTS

Database packages let you make attractive management reports from a table's records or from a query's output. If you are making a report from a table, the Access report generator looks up the data in the table and puts it into report format. If you are making a report from a query's output, Access runs the query in the background (you do not control it or see it happen) and then puts the output in report format.

There are three ways to make a report. One is to hand-craft the report from scratch in the Design View. Because that is a tedious process, it is not shown in this tutorial. The second way is to use the Report Wizard, during which Access leads you through a menu-driven construction. That method is shown in this tutorial. The third way is to start in the Wizard and then use the Design View to tailor what the Wizard produces. That method also is shown in this tutorial.

Creating a Grouped Report

This tutorial assumes that you already know how to use the Wizard to make a basic ungrouped report. This section of the tutorial teaches you how to make a grouped report. (If you don't know how to make an ungrouped report, you can learn by following the first example in the upcoming section.)

AT THE KEYBOARD

Suppose you want to make a report out of the HOURS WORKED table. Choose the Create tab, Report Wizard in the Reports group. Select the HOURS WORKED table from the drop-down menu as the report basis. Select all of the fields (using the **>>** button), as shown in Figure B-63.

Click **Next**. Then tell Access you want to group on Week # by double-clicking that field name. This grouping step is shown in Figure B-64.

Click **Next**. You'll see a screen, similar to the one in Figure B-65, for Sorting and Summary Options.

Because you chose a grouping field, Access will let you decide whether you want to see group subtotals and/or report grand totals. If you choose that option, all numeric fields could be added. In this example, group subtotals are for total hours in each week. Assume you *do* want the total of hours by week. Click **Summary Options**. You'll get a screen similar to the one in Figure B-66.

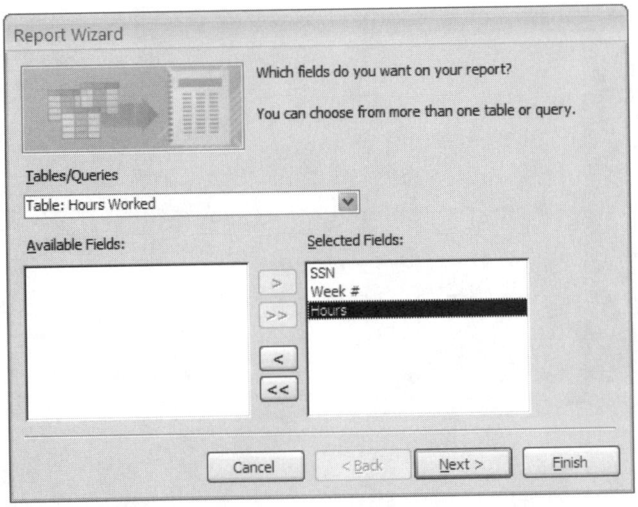

FIGURE B-63 Field selection step in the Report Wizard

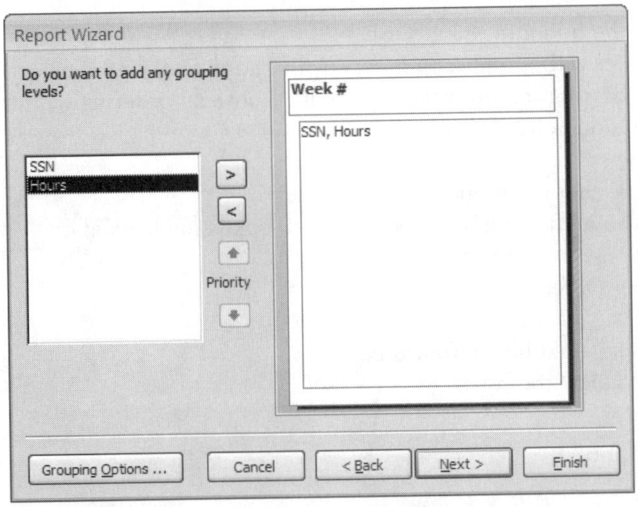

FIGURE B-64 Grouping step in the Report Wizard

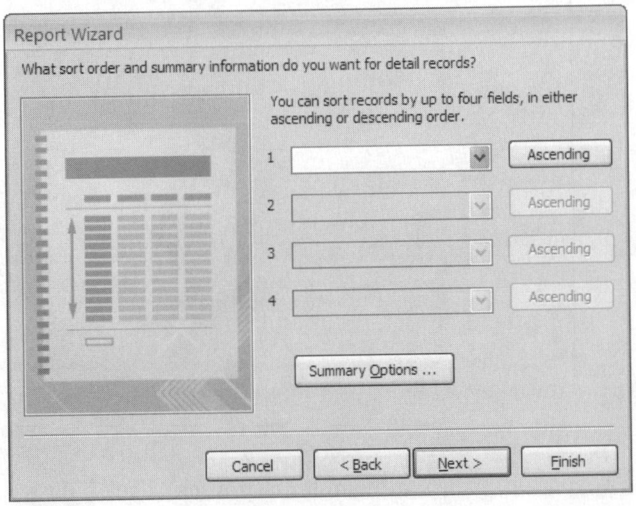

FIGURE B-65 Sorting and Summary Options step in the Report Wizard

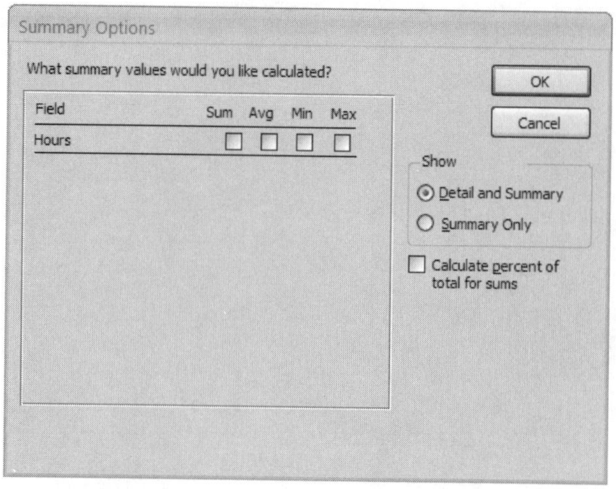

FIGURE B-66 Summary Options in the Report Wizard

Next, follow these steps:

1. Click the Sum box for Hours (to sum the hours in the group).
2. Click **Detail and Summary**. (Detail equates with "group"; Summary, with "grand total for the report.")
3. Click **OK**. This takes you back to the Sorting screen, where you can choose an ordering within the group if desired. (In this case, you don't choose one.)
4. Click **Next** to continue.
5. In the Layout screen (not shown here), choose Stepped and Portrait.
6. Make sure the "Adjust the field width so all fields fit on a page" check box is unchecked.
7. Click **Next**.
8. In the Style screen (not shown), accept Office.
9. Click **Next**.
10. Provide a title—Hours Worked by Week would be appropriate.
11. Select the Preview the report button to view the report.
12. Click **Finish**.

Your report will look like that shown in Figure B-67.

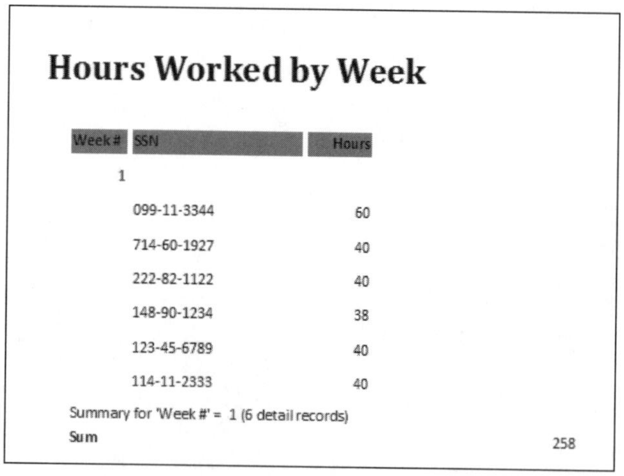

Hours Worked by Week

Week #	SSN	Hours
1		
	099-11-3344	60
	714-60-1927	40
	222-82-1122	40
	148-90-1234	38
	123-45-6789	40
	114-11-2333	40

Summary for 'Week #' = 1 (6 detail records)

Sum 258

FIGURE B-67 Hours Worked by Week report

Notice that data is grouped by weeks, with Week 1 on top, then a subtotal for that week. Week 2 data is next, then a grand total (which you can scroll down to see). The subtotal is labeled "Sum," which is not very descriptive. That can be changed later in the Design View. Also, there is the apparently useless "Summary for Week . . ." line, which can be deleted later in the Design View as well. Click the *x* in the upper-right corner of your report window corresponding to "Close Hours Worked by Week." Note that the Navigation Pane on the left side of your screen contains the report shown in Figure B-68.

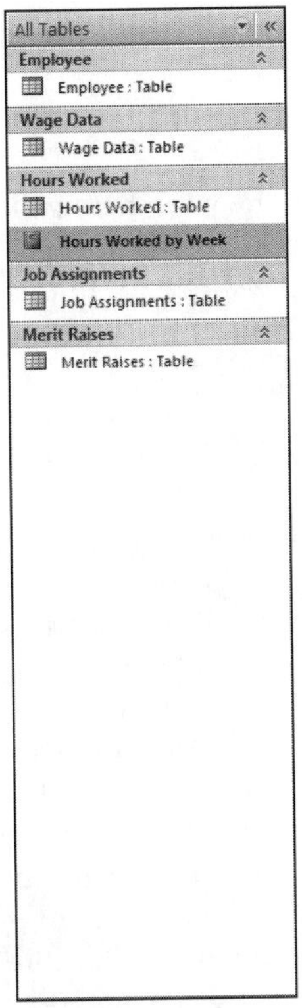

FIGURE B-68   Navigation Pane with report

To edit the report in the Design View, right-click the report title and choose Design view. You will see a complex (and intimidating) screen, similar to the one shown in Figure B-69.

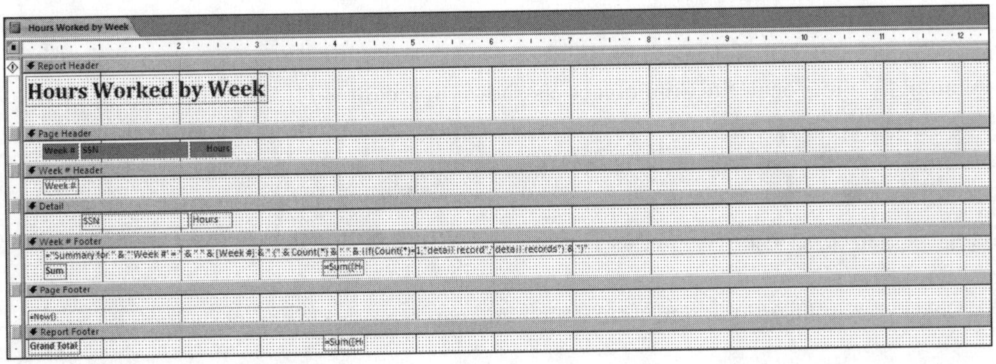

FIGURE B-69   Report design screen

The organization of the screen is hierarchical. At the top is the Report level. The next level down (within a report) is the Page level. The next levels down (within a page) are for any data groupings you have specified.

If you told Access to make group (summary) totals, your report would have a Report Header area and end with a Grand Total in the Report Footer. The report header is usually just the title you specified.

A page also has a header, which is usually just the names of the fields you told Access to put in the report (here, Week #, SSN, and Hours fields). Sometimes the page number is put in by default.

Groupings of data are more complex. There is a header for the group. In this case, the *value* of the Week # will be the header, so the values shown will be 1 and 2. Those headers indicate that there is a group of data for the first week, then a group of data for the second week. Within each data grouping is the other "detail" that you've requested. In this case, there will be data for each SSN and the related hours.

Each Week # gets a footer, which is a labeled sum. Recall that you asked for the footer to be shown (i.e., Detail and Summary were requested). The Week # Footer is composed of three elements:

1. The line that starts =Summary for . . .
2. The Sum label
3. The adjacent expression =Sum(Hours)

The line beneath the Week # Footer will be printed unless you eliminate it. Similarly, the word *Sum* will be printed as the subtotal label unless you eliminate it. The =Sum(Hours) is an expression that tells Access to add the quantity *for the header in question* and to put that number into the report as the subtotal. (In this example, that number would be the sum of hours by Week #.)

Each report also gets a footer—the grand total (in this case, of hours) for the report.

If you look closely, each of the detail items appears to be doubly inserted in the design. For example, in Figure B-69, you will see the notation for SSN twice, once in the Page Header band and then again in the Detail band. Hours are treated similarly.

Those items will not, however, be printed twice, because each detail item in the report is an object. In Access, an object is denoted by both a label and its value. In other words, there is a representation of the name of the object, which is the boldfaced name itself (in this example, *SSN* in the Page Header band), and there is a representation in less bold type of the value of the object (corresponding to *SSN* in the Detail band).

Sometimes the Report Wizard is arbitrary about where it puts labels and data. If you do not like where the Wizard puts data, you can move the objects containing the data around in the Design View. Also, you can click and drag within the band or across bands. Often a box will be too small to allow the full numerical values for that object to show. When that happens, select the box and click one of the sides to stretch it. Doing so will allow full values to show. At other times, an object's box will be very long. When that happens, you can click the box, resize it, then drag right or left in its panel to reposition the output.

Suppose you do *not* want the Summary for . . . line to appear in the report. Also suppose you would like different subtotal and grand total labels. The Summary for . . . line is an object you can activate by clicking. Do that now. "Handles" (little squares) should appear around its edges, as shown in Figure B-70.

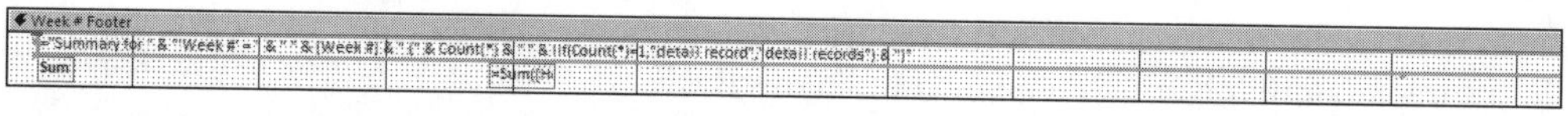

FIGURE B-70   Selecting an object in the Report Design View

Press the Delete key to get rid of the selected object.
To change the subtotal heading, click the Sum object, as shown in Figure B-71.

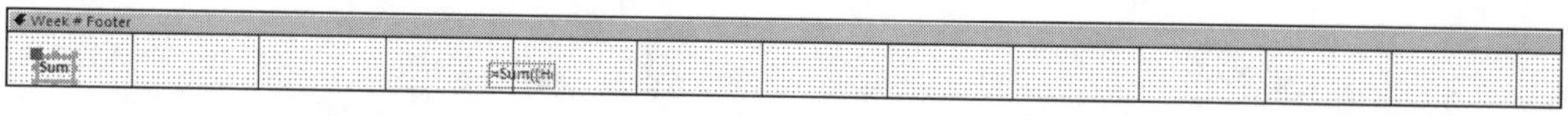

FIGURE B-71   Selecting the Sum object in the Report Design View

Click again. That gives you an insertion point from which you can type, as shown in Figure B-72.

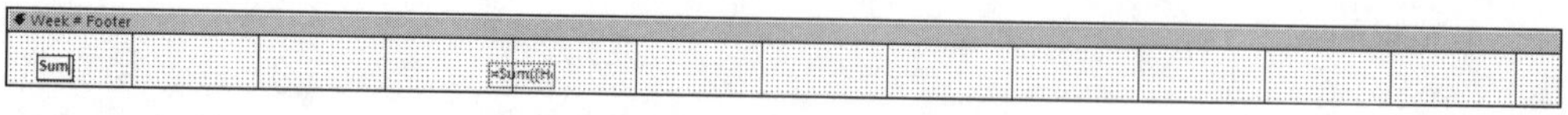

FIGURE B-72   Typing an object in the Report Design View

Change the label to Sum of Hours for Week and press Enter or click somewhere else in the report to deactivate the object. Your screen should resemble the one shown in Figure B-73.
You can change the Grand Total in the same way.
In the Design tab, click the View button on the Views group. You should see a report similar to that in Figure B-74 (the top part is shown).

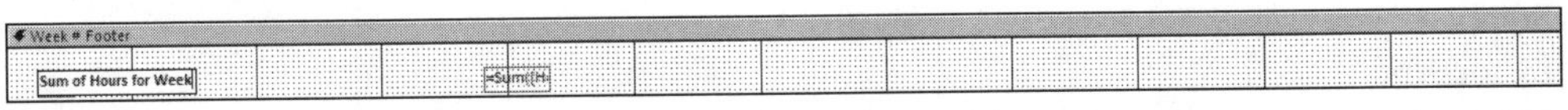

FIGURE B-73   Changing a label in the Report Design View

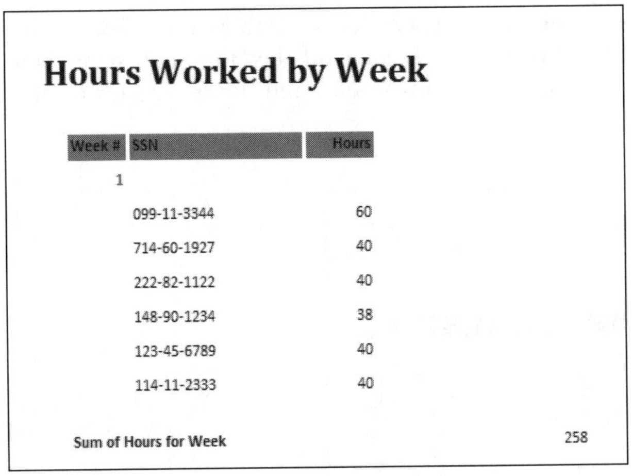

FIGURE B-74   Hours Worked by Week report

Notice that the data is grouped by week number (data for Week 1 is shown) and subtotaled for that week. The report also has a grand total at the bottom.

Moving Fields in the Design View

If you want to group on more than one field in the Report Wizard, the report will have an odd "staircase" look, display repeated data, or it will have both features. Next, you will learn how to overcome that effect in the Design View.

Suppose you make a query showing an employee's last name, first name, street address, zip code, week #, and hours worked. Then you make a report from that query, grouping on last name and first name only. See Figure B-75.

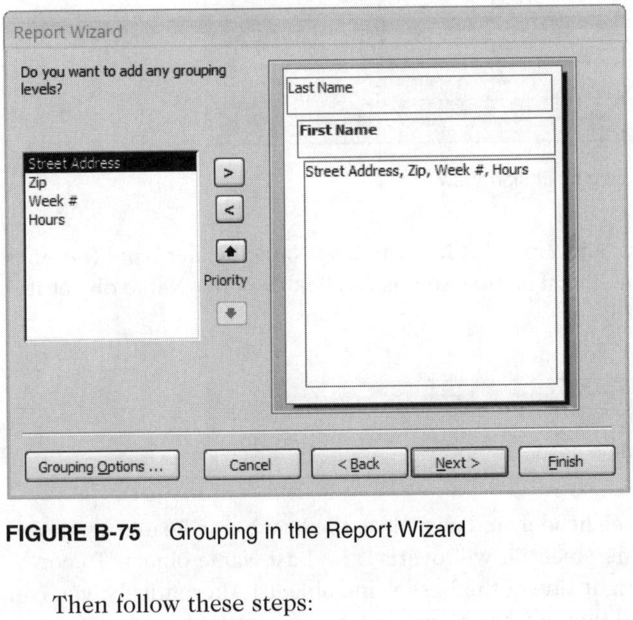

FIGURE B-75   Grouping in the Report Wizard

Then follow these steps:

1. Click **Next**.
2. Do not sum anything in Summary Options. Click **Next**.

3. Select Stepped.
4. Select Landscape.
5. Click the check mark by *Adjust the field width so all fields fit on a page*. Click **Next**.
6. Select Office. Click **Next**.
7. Type a title (Hours Worked by Employees). Click **Finish**.

As you preview the report, you will notice repeating data. In the report displayed in Figure B-76, notice that the Zip data and Street Address data are repeating and that both are shown below the First Name data, which is below the Last Name data—hence, the staircase effect. (The fields Week # and Hours are shown subordinate to Last Name and First Name, as desired.)

Hours Worked by Employees

Last Name	First Name	Street Address	Zip	Week #	Hours
Brady					
	Joe				
		2 Main St	21921	2	55
		2 Main St	21921	1	60

FIGURE B-76   Hours Worked by Employees grouped report

Suppose you want the last name, first name, street address, and zip code to appear on the same line. If so, take the report into the Design View for editing. From the Navigation Pane on the left side of your screen, right-click the Hours Worked by Employees report and choose Design View. At this point, the headers look like those shown in Figure B-77.

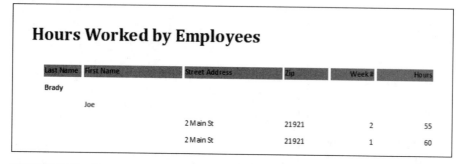

FIGURE B-77   Hours Worked by Employees report displayed in Design View

Your goal is to get the First Name, Street Address, and Zip fields into the Last Name header band (*not* into the Page Header band) so they will print on the same line. The first step is to click the First Name object in the First Name Header band, as shown in Figure B-78.

FIGURE B-78   Selecting First Name object in the First Name header

Right-click, cut, place your cursor in the Last Name header, and paste into the Last Name header, as shown in Figure B-79. (*Note:* When you paste the First Name object, it will overlay the Last Name object. To correct that, simply move the First Name object to the right side of the Last Name object.) Alternatively, you can right-click the First Name object, choose Layout, and then choose Move Up a Section.

Now cut the Street Address object and paste it into the Last Name header, as shown in Figure B-80.

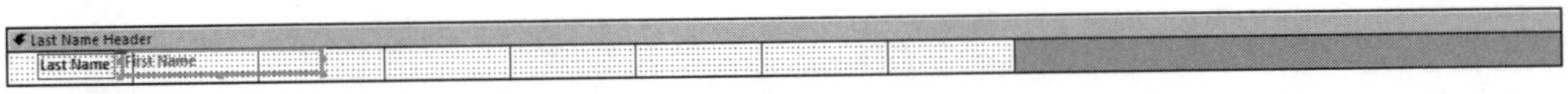

FIGURE B-79   Moving the First Name object to the Last Name header

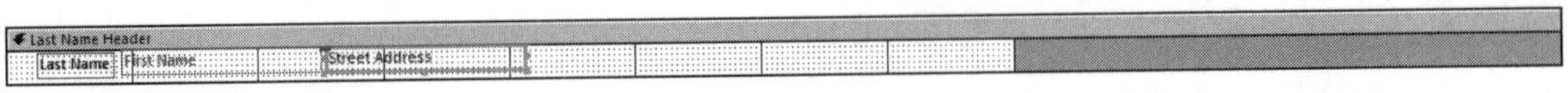

FIGURE B-80   Moving the Street Address object to the Last Name header

Do the same for the Zip object. Now correct the spacing and the Page Header by choosing the Layout View from the View group on the Design tab. Click **Hours** and note that the entire column is selected (your heading and data). Drag the column to the right side of the report. At this point, your screen should look like that shown in Figure B-81.

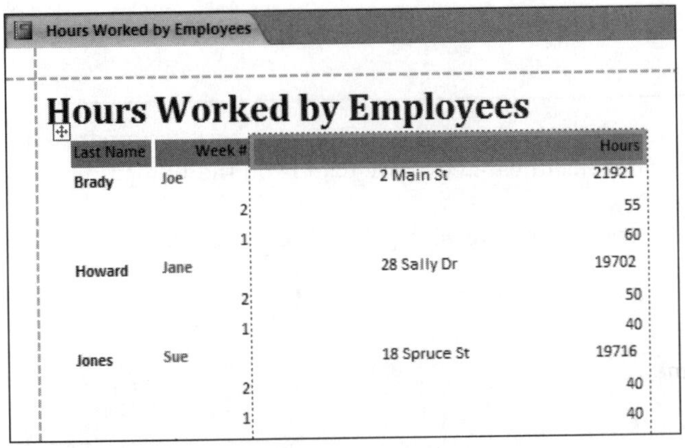

FIGURE B-81   Adjusting report layout

Drag the left side of the Hours column to the right; the Week # heading and data will follow your cursor to the right. Your report should now resemble the one shown in Figure B-82.

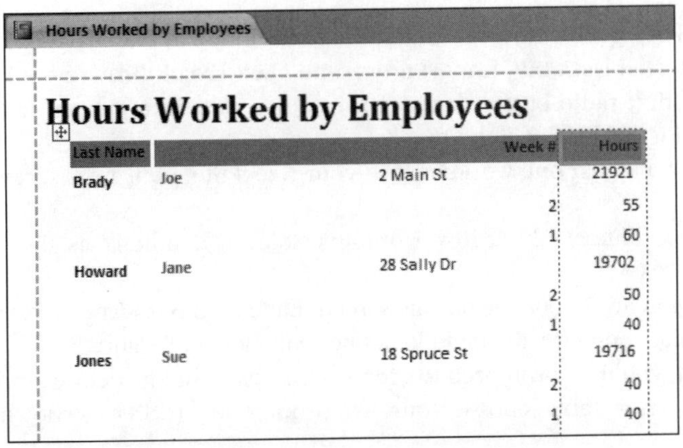

FIGURE B-82   Adjusting report layout, continued

Finally, adjust the headings by dragging the left edge of the Week # to the right and adding more descriptions such as First Name, Address, and Zip to the Last Name header. This final adjustment will result in a report similar to that shown in Figure B-83. (*Note:* This is the Report View.)

Hours Worked by Employees

Last Name, First Name, Address and Zip				Week #	Hours
Brady	Joe	2 Main St	21921		
				2	55
				1	60
Howard	Jane	28 Sally Dr	19702		
				2	50
				1	40
Jones	Sue	18 Spruce St	19716		
				2	40
				1	40

FIGURE B-83   Hours Worked by Employees report

IMPORTING DATA

Text or spreadsheet data is easily imported into Access. In business, it is often necessary to import data because companies use disparate systems. Assume your healthcare coverage data is on the human resources manager's computer in an Excel spreadsheet.

Open the software application Microsoft Excel. Create a spreadsheet in Excel using the data shown in Figure B-84.

	A	B	C
1	SSN	Provider	Level
2	114-11-2333	BlueCross	family
3	123-45-6789	BlueCross	family
4	148-90-1234	Coventry	spouse
5	222-82-1122	None	none
6	714-60-1927	Coventry	single
7	Your SSN	BlueCross	single

FIGURE B-84   Excel data

Save the file; then close it. Now you can easily import that spreadsheet data into a new table in Access. With your **Employee** database open, go to the External Data tab, Import group and click **Excel**. Browse to find the Excel file you just created and make sure the first radio button (Import the source data into a new table in the current database) is selected, as shown in Figure B-85. Click **OK**.

Choose the correct worksheet. Assuming you have just one worksheet in your Excel file, your next screen should look like that shown in Figure B-86.

Choose Next and make sure you select the box that says First Row Contains Column Headings, as shown in Figure B-87.

Choose Next. Accept the default for each field you are importing on this screen. Each field is assigned a text data type, which, in this case, is correct for this table. Your screen should look like that shown in Figure B-88.

Choose Next. In the next screen of the Wizard, you'll be prompted to create an index—that is, define a primary key. Because you will be storing your data in a new table, choose your own primary key (SSN), as shown in Figure B-89.

Continue through the Wizard, giving your table an appropriate name. After importing the table, take a look at it and its design. (Highlight the Table option and use the Design button.) Note the width of each field (very large). Adjust the field properties as needed.

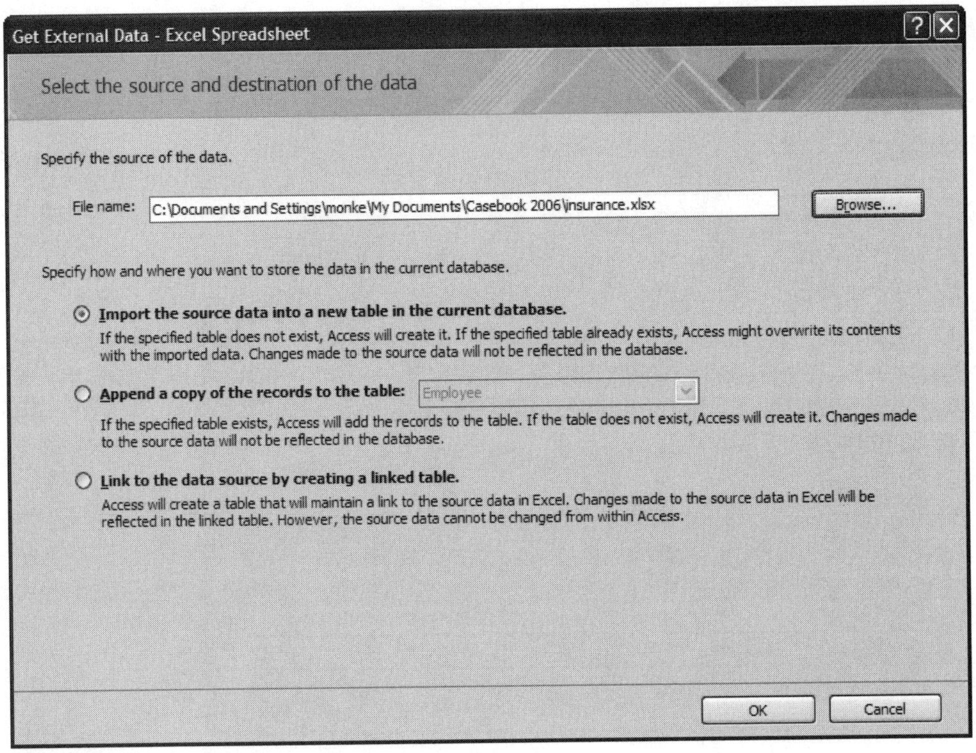

FIGURE B-85   Importing Excel data into a new table

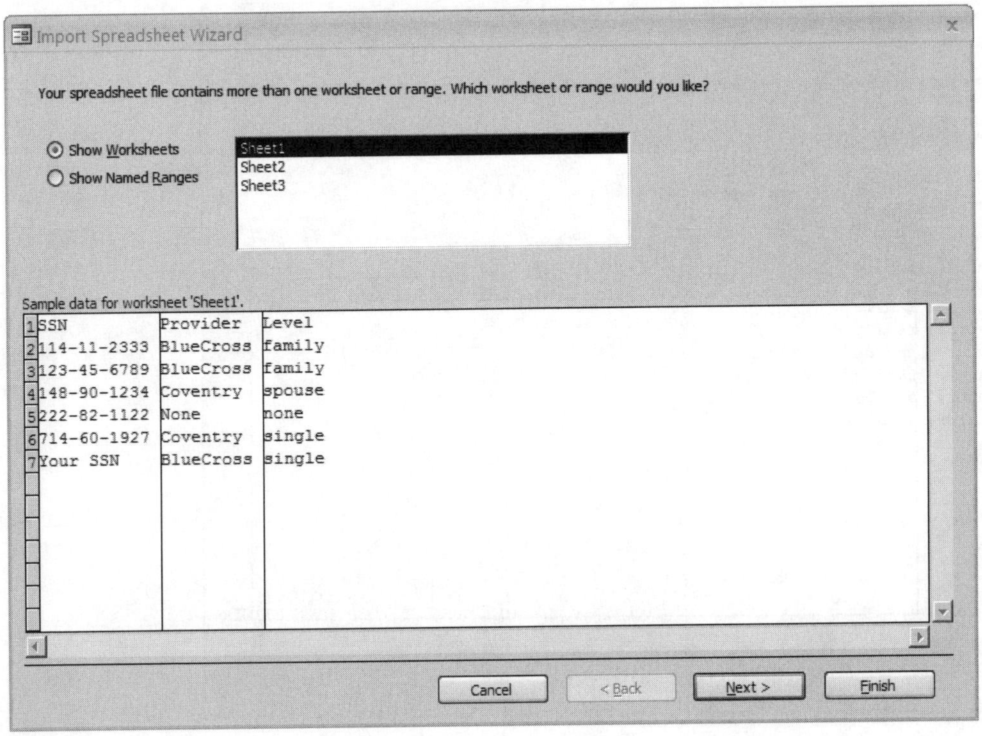

FIGURE B-86   First screen in the Import Spreadsheet Wizard

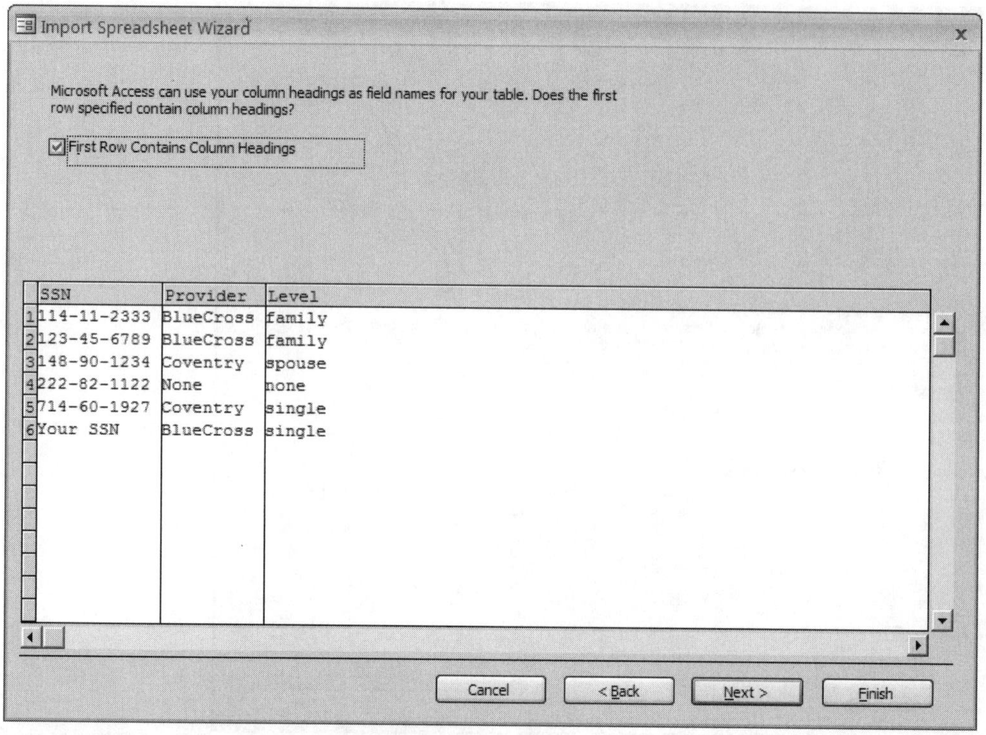

Import Spreadsheet Wizard   x

Microsoft Access can use your column headings as field names for your table. Does the first row specified contain column headings?

☑ First Row Contains Column Headings

	SSN	Provider	Level
1	114-11-2333	BlueCross	family
2	123-45-6789	BlueCross	family
3	148-90-1234	Coventry	spouse
4	222-82-1122	None	none
5	714-60-1927	Coventry	single
6	Your SSN	BlueCross	single

Cancel   < Back   Next >   Finish

FIGURE B-87   Choosing column headings in the Import Spreadsheet Wizard

Import Spreadsheet Wizard   x

You can specify information about each of the fields you are importing. Select fields in the area below. You can then modify field information in the 'Field Options' area.

Field Options

Field Name: SSN   Data Type: Text
Indexed: No   ☐ Do not import field (Skip)

	SSN	Provider	Level
1	114-11-2333	BlueCross	family
2	123-45-6789	BlueCross	family
3	148-90-1234	Coventry	spouse
4	222-82-1122	None	none
5	714-60-1927	Coventry	single
6	Your SSN	BlueCross	single

Cancel   < Back   Next >   Finish

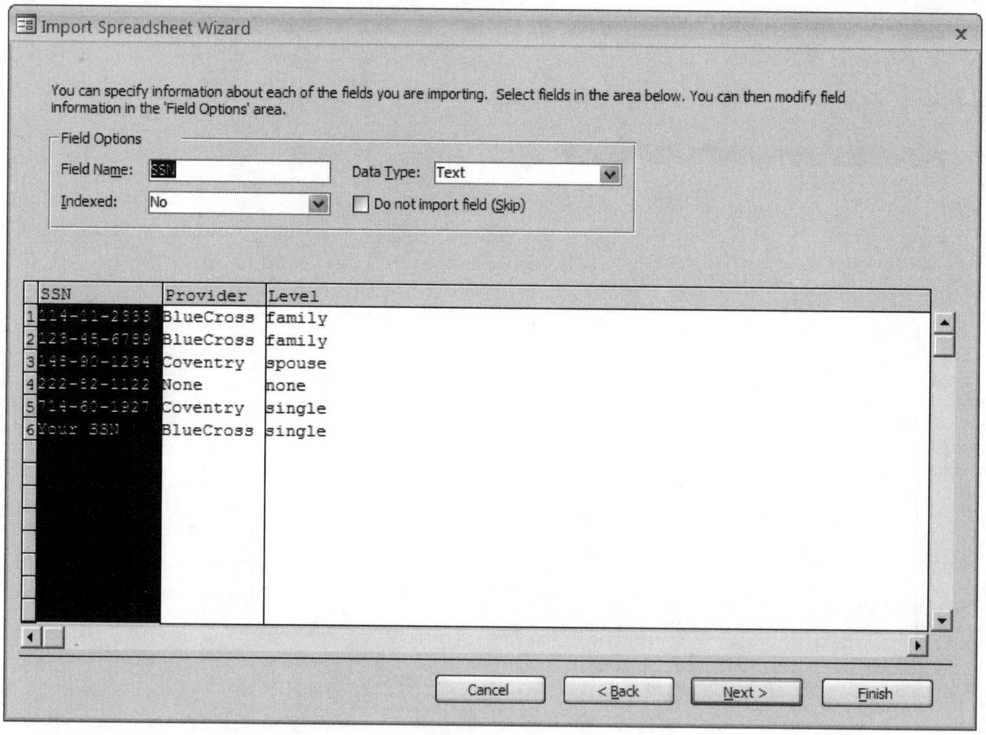

FIGURE B-88   Choosing data type for each field in the Import Spreadsheet Wizard

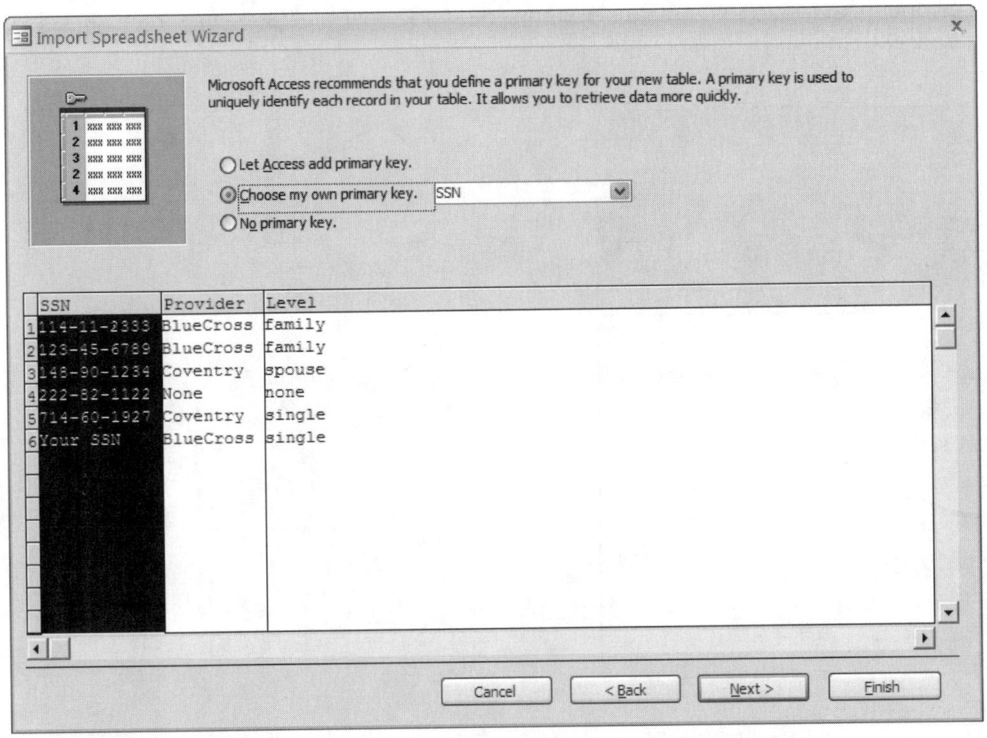

FIGURE B-89 Choosing a primary key field in the Import Spreadsheet Wizard

MAKING FORMS

Forms simplify the process of adding new records to a table. The Form Wizard is easy to use and can be applied to a single table or to multiple tables.

When you base a form on one table, you simply identify that table when you are in the Form Wizard setup. The form will then have all of the fields from that table and only those fields. When data is entered into the form, a complete new record is automatically added to the table. Forms with two tables are discussed next.

Making Forms with Subforms

You also can make a form that contains a subform, which can be useful when you need to create a form that is based on two (or more) tables. Return to the example Employee database to see how forms and subforms would be particularly handy for viewing all of the hours that each employee worked each week. Suppose you want to show all of the fields from the EMPLOYEE table; you also want to show the hours each employee worked (in other words, include all fields from the HOURS WORKED table as well).

Creating the Form and Subform

To create the form and subform, first create a simple one-table form using the Form Wizard on the EMPLOYEE table. Follow these steps:

1. In the Create tab, Forms group, choose More Forms, Form Wizard.
2. Make sure the EMPLOYEE table is selected under the drop-down menu of Tables/Queries.
3. Select all Available Fields by clicking the right double-arrow button.
4. Select Next.
5. Select Columnar layout.
6. Select Next.
7. Select Office Style.
8. Select Next.
9. When asked "What title do you want for your form?", type *Employee Hours*.
10. Select Finish.

After the form is complete, in the Design tab, Views group, choose Design View so your screen looks like the one shown in Figure B-90.

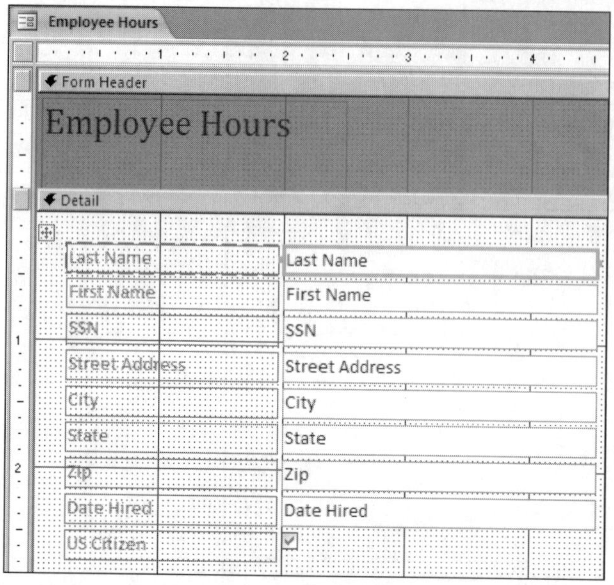

FIGURE B-90   The Employee Hours form

In the Design tab, Controls group, locate the Subform/Subreport button, which is above the letter *C* in *Controls*, as shown in Figure B-91.

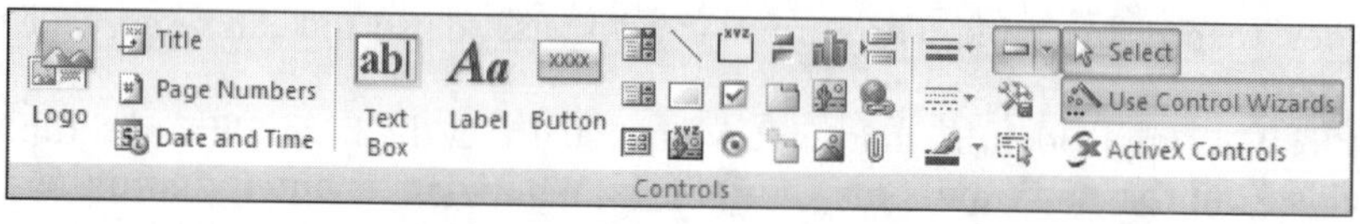

FIGURE B-91   The Controls group

Click the Subform/Subreport button (sixth row, button on right) and using your cursor, drag a small section next to the State, Zip, Date Hired, and US Citizen fields in your form design. As you lift your cursor, the Subform Wizard will appear, as shown in Figure B-92. (*Note:* You may need to stretch the area of the form before placing the subform.)

Follow these steps to create data in the subform:

1. Select the button Use existing Tables and Queries.
2. Select Next.
3. Under Tables/Queries, choose the HOURS WORKED table and bring all fields into the Selected Fields box by clicking the right double-arrow button.
4. Select Next.
5. Select the Choose from a list radio button.
6. Select Next.
7. Use the default subform name.
8. Select Finish.

Now you need to adjust the design so the data of all of the fields are visible. Go to the Datasheet View and click through the various records to see how the subform data changes. Your final form should resemble the one shown in Figure B-93.

FIGURE B-92   The Subform Wizard

Employee Hours

Last Name	Brady
First Name	Joe
SSN	099-11-3344
Street Address	2 Main St
City	Elkton
State	MD
Zip	21921
Date Hired	1/27/2007
US Citizen	☑

Hours Worked subform

SSN	Week #	Hours
099-11-3344	1	60
099-11-3344	2	55
099-11-3344		

Record: 1 of 2   No Filter   Search

FIGURE B-93   The Employee Hours form with the Hours Worked subform

CREATING A CUSTOM NAVIGATION PANE

If you want someone who knows nothing about Access to be able to run your database, you can create a customized area, known as a Navigation Pane, to simplify the work. A Navigation Pane provides a simple, user-friendly interface that has groupings of objects that the user can click to perform certain tasks. For example, you can design a custom Navigation Pane with two groupings: one for the Forms and one for the Reports. Your finished product will show the objects within each grouping along the left side of the screen. The user can open each object by simply clicking on it.

To design that custom Navigation Pane, right-click the top of the Navigation Pane and choose Navigation Options.

The Navigation Options dialog box will open, as shown in Figure B-94.

Click the Add Item button, which appears below the Categories column, as shown in Figure B-95.

Change the Custom Group 1 name to Easy Navigation and press Enter.

With the Easy Navigation still selected, click the Add Group button, which appears below the right column, Groups for "Easy Navigation." Add a group for Forms and a group for Reports. Leave the Unassigned Objects group as is. Your screen should look like the one shown in Figure B-96.

Choose OK. Your screen should look like the one shown in Figure B-97.

Now you need to add objects to your new groups. Choose Easy Navigation from the top of the Navigation Pane.

Drag each form in the Unassigned Objects group to the Forms group and drag each report to the Reports group. When you are finished, you can hide the Unassigned Objects group by clicking the double arrow in the Unassigned Objects bar. Your completed Easy Navigation Pane should look like that shown in Figure B-98.

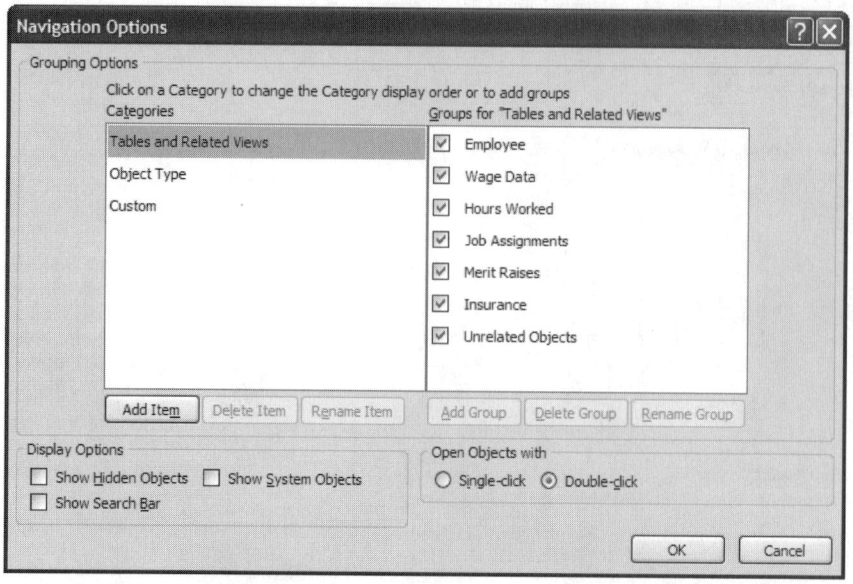

FIGURE B-94   The Navigation Options dialog box

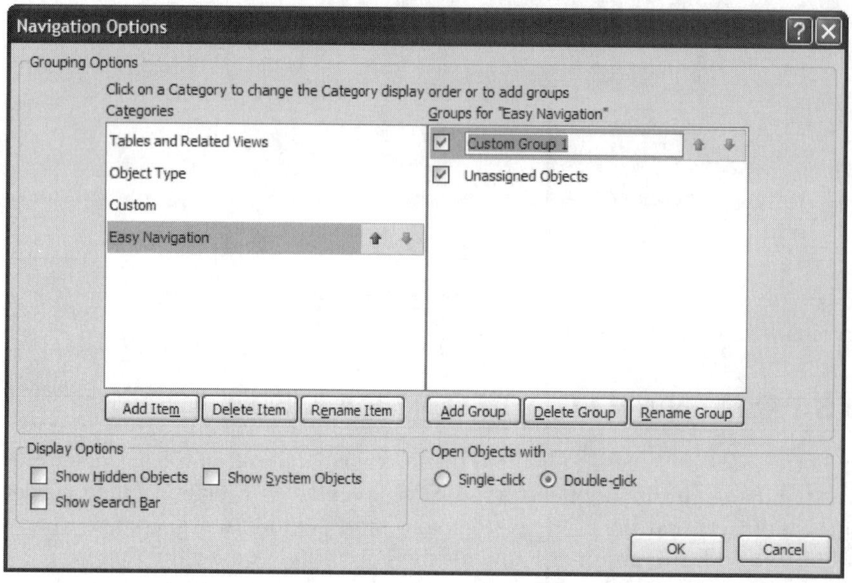

FIGURE B-95   Add Item in the Navigation Options dialog box

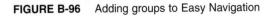

FIGURE B-96   Adding groups to Easy Navigation

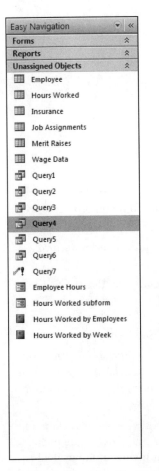

FIGURE B-97   Easy Navigation without objects

FIGURE B-98   Easy Navigation completed

TROUBLESHOOTING COMMON PROBLEMS

Access beginners (and veterans) sometimes create databases that have problems. Some common problems are described below, along with their causes and corrections.

1. *"I saved my database file, but I can't find it on my computer or external secondary storage medium! Where is it?"*

    You saved your file to some fixed disk or some place other than My Documents. Use the Search option of the Windows Start button. Search for all files ending in .accdb (search for *.accdb). If you did save the file, it is on the hard drive (C:\) or on some network drive. (Your site assistant can tell you the drive designators.)

2. *"What is a 'duplicate key field value'? I'm trying to enter records into my Sales table. The first record was for a sale of product X to customer 101, and I was able to enter that one. But when I try to enter a second sale for customer #101, Access tells me I already have a record with that key field value. Am I allowed to enter only one sale per customer?"*

    Your primary key field needs work. You may need a compound primary key—customer number and some other field(s). In this case, customer number, product number, and date of sale might provide a unique combination of values; or you might consider using an invoice number field as a key.

3. *"My query says 'Enter Parameter Value' when I run it. What is that?"*

    This symptom, 99 times out of 100, indicates that you have an expression in a Criteria or a calculated field in which *you have misspelled a field name*. Access is very fussy about spelling. For example, Access is case-sensitive. The program is also "space-sensitive," which means when you put a space in a field name when you define the table, you must put a space in the field name when you reference it in a query expression. Fix the typo in the query expression.

4. *"I'm getting a fantastic number of rows in my query output—many times more than I need. Most of the rows are duplicates!"*

This symptom is usually caused by a failure to link all of the tables you brought into the top half of the query generator. The solution is to use the manual click-and-drag method to link the common fields between tables. (Spelling of the field names is irrelevant because the link fields need not be spelled the same.)

5. *"For the most part, my query output is what I expected; but I am getting one or two duplicate rows or not enough rows."*

You may have linked too many fields between tables. Usually, only a single link is needed between two tables. It's unnecessary to link each common field in all combinations of tables; usually, it's enough to link the primary keys. A layperson's explanation for why overlinking causes problems is that excess linking causes Access to "overthink" the problem and repeat itself in its answer. On the other hand, you might be using too many tables in the query design. For example, you brought in a table, linked it on a common field with some other table, but then did not use the table—in other words, you brought down none of its fields and/or you used none of its fields in query expressions. In this case, if you were to get rid of the table, the query would still work. Try doing the following to see whether the few duplicate rows disappear: Click the unneeded table's header in the top of the QBE area and press the Delete key.

6. *"I expected six rows in my query output, but I only got five. What happened to the other one?"*

Usually, this indicates a data entry error in your tables. When you link the proper tables and fields to make the query, remember that the linking operation joins records from the tables *on common values* (*equal* values in the two tables). For example, if a primary key in one table has the value "123", the primary key or the linking field in the other table should be the same to allow linking. Note that the text string "123" is not the same as the text string "123 "—the space in the second string is considered a character too. Access does not see unequal values as an error. Instead, Access moves on to consider the rest of the records in the table for linking. Solution: Look at the values entered into the linked fields in each table and fix any data entry errors.

7. *"I linked fields correctly in a query, but I'm getting the empty set in the output. All I get are the field name headings!"*

You probably have zero common (equal) values in the linked fields. For example, suppose you are linking on Part Number (which you declared as text). In one field, you have part numbers "001", "002", and "003"; and in the other table, you have part numbers "0001", "0002", and "0003". Your tables have no common values, which means no records are selected for output. You'll have to change the values in one of the tables.

8. *"I'm trying to count the number of today's sales orders. A Totals query is called for. Sales are denoted by an invoice number, and I made that a text field in the table design. However, when I ask the Totals query to 'Sum' the number of invoice numbers, Access tells me I cannot add them up! What is the problem?"*

Text variables are words! You cannot add words, but you can count them. Use the Count Totals operator (not the Sum operator) to count the number of sales, each being denoted by an invoice number.

9. *"I'm doing time arithmetic in a calculated field expression. I subtracted the Time In from the Time Out and got a decimal number! I expected eight hours, and I got the number .33333. Why?"*

[Time Out] – [Time In] yields the decimal percentage of a 24-hour day. In your case, eight hours is one-third of a day. You must complete the expression by multiplying by 24: ([Time Out] – [Time In]) * 24. Don't forget the parentheses.

10. *"I formatted a calculated field for Currency in the query generator, and the values did show as currency in the query output; however, the report based on the query output does not show the dollar sign in its output. What happened?"*

Go into the report Design View. A box in one of the panels represents the calculated field's value. Click the box and drag to widen it. That should give Access enough room to show the dollar sign as well as the number in the output.

11. *"I told the Report Wizard to fit all of my output to one page. It does print to one page, but some of the data is missing. What happened?"*

Access fits all the output on one page by *leaving data out*. If you can stand to see the output on more than one page, click off the Fit to a Page option in the Wizard. One way to tighten output is to go into the Design View and remove space from each of the boxes representing output values and labels. Access usually provides more space than needed.

12. *"I grouped on three fields in the Report Wizard, and the Wizard prints the output in a staircase fashion. I want the grouping fields to be on one line. How can I do that?"*

Make adjustments in the Design View and the Layout View. See the Reports section of this tutorial for instruction on how to make those adjustments.

13. *"When I create an Update query, Access tells me that zero rows are updating or more rows are updating than I want. What is wrong?"*

If your Update query is not set up correctly (for example, if the tables are not joined properly), Access will try not to update anything or will update all of the records. Check the query, make corrections, and run it again.

14. *"After making a Totals Query with a Sum in the Group By row and saving that query, when I go back to it, the Sum field says Expression and Sum is put in the field name box. Is that wrong?"*

Access sometimes changes that particular statistic when the query is saved. The data remains the same, and you can be assured your query is correct.

15. *"I am not able to run my Update Query, but I know it is set up correctly. What is wrong?"*

Check the Security Content of the database by clicking the Security Content button. You may need to enable certain actions.

PRELIMINARY CASE: FURNITURE RENTALS

Setting up a Relational Database to Create Tables, Forms, Queries, and Reports

PREVIEW

In this case, you'll create a relational database for a small company that rents furniture to college students. First, you'll create four tables and populate them with data. Next, you'll create a form and subform for recording rentals, along with four queries: two Select queries, a Totals query, and a query with a calculated field. Finally, you'll create a report from the last query that displays the rental money due.

PREPARATION

- Before attempting this case, you should have some experience using Microsoft Access.
- Complete any part of Access Tutorial B that your instructor assigns, or refer to the tutorial as necessary.

BACKGROUND

Students at universities all over the country move into accommodations and then need certain items to help them live as comfortably as possible. Furniture Rental, Inc. (FRI) steps in where apartments leave off and rents sturdy furniture to college students for the academic year. This allows students to have good desks, chairs, lamps, sofas, and other furniture without buying the items. FRI has been able to strike deals with universities that allows the company to put the rental costs on the student's bursar bill. This method avoids any direct billing from FRI to the student and always results in full collection of the rental fees.

The company is growing rapidly and needs your help. An owner of FRI heard that you are proficient in Microsoft Access. You have been hired to finish the database project and you begin the job with the tables already designed for you. The database has four tables:

- The CUSTOMERS table keeps track of customer information such as ID, name, address, telephone number, and e-mail address.
- The INVENTORY table keeps track of all the items available for rental—in particular, the item ID, its description, and price per month. Assume that an unlimited supply of furniture is available for rental. This is a reasonable assumption because FRI has 10 large warehouses across the country and items are always available from at least one nearby warehouse.
- The RENTALS table records an identification number for each rental, the customer who rented the item(s), the date they were rented, and the date they were returned.
- The RENTAL LINE ITEM table records the individual item IDs rented within each rental ID, along with the quantity of each item rented.

The owners of FRI have a few requirements for information output that they would like to see in the database beyond simple data recording. First, they want to be able to record a rental order complete with details of what customers rented. This task can be accomplished with a form and subform.

In addition, the owners want the database to answer some questions. First, they want to be able to see which customers rented specific items. For example, the kitchen chairs that FRI has been renting are of very poor quality and often break. After complaints to the manufacturer, FRI has been able to replace all defective chairs.

The company needs a listing of renters who have defective chairs so it can contact the renters and replace the chairs quickly. Next, the company wants to be able to send reminders to students who have not returned any rental items by December 15. This reminder is sent to make sure that students remember to return all rental items, even if they are graduating mid-year or simply moving on at the end of the fall semester.

Of course, the company needs to keep track of the most popular furniture for rental so it can plan future marketing promotions and warehouse stocking. Also, the company would like a report that calculates and then shows all the rental money billed to the student for the fall semester.

ASSIGNMENT 1: CREATING TABLES

Use Microsoft Access to create the tables with the fields shown in Figures 1-1 through 1-4; these tables were discussed in the Background section. Populate the database tables as shown. Add your name to the CUSTOMERS table, assign yourself a student ID of 110, and fill in your address, phone number, and e-mail address.

Customer ID	Last Name	First Name	Address	City	State	Zip	Telephone	Email	Add New Field
101	Smith	John	34 Redback Road	Austin	TX	78701	(512)998-0675	Jsmith@pcp.net	
102	Chavez	Mary	9 Elm Arch Road	Woods Hole	MA	02543	(508)443-3212	Jmary@comcast.net	
103	Trainer	Lisa	9000 Rt 40, Apt 25-A	San Francisco	CA	94102	(415)217-9065	readinglisa@aol.com	
104	O'Hara	Heather	20 Cheswold Blvd	Cheyenne	WY	82022	(307)887-0176	russ101@aol.com	
105	Thompson	Cynthia	31 Spectrum Drive	Portland	OR	97086	(541)443-1123	loohoo@zoom.net	
106	Isaacs	Irving	10010 N. Barrett Dr	River Edge	NJ	07661	(201)584-1028	irv@aol.com	
107	Downing	Charlie	200 Mac Duff Rd	Provo	UT	84601	(801)912-6564	charlieboy@comcast	
108	DeVivo	Paul	625 Dawson Rd	Baton Rouge	LA	70801	(985)453-0943	tanz@zoom.net	
109	Bevans	Beverly	24 Raway Rd	Livonia	MI	48150	(734)452-0914	bev@comcast.net	

FIGURE 1-1 The CUSTOMERS table

Item ID	Description	Price Per Mo	Add New Field
C-101	Desk Chair	$15.00	
C-102	Kitchen Chair	$12.00	
D-101	Single Dresser	$30.00	
D-102	Double Dresse	$35.00	
D-103	Triple Dresser	$40.00	
D-104	Desk	$28.00	
L-101	Desk Lamp	$8.00	
L-102	Table Lamp	$10.00	
L-103	Pole Lamp	$12.00	
S-101	Double Sofa	$25.00	
S-102	Triple Sofa	$30.00	
S-103	Sofa Bed	$35.00	
T-101	Kitchen Table	$20.00	
T-102	Coffee Table	$12.00	
T-103	TV Table	$10.00	

FIGURE 1-2 The INVENTORY table

Add your own data to the RENTALS and RENTAL LINE ITEM tables, as shown in the following two figures.

Rentals				
Rental Number	Customer Number	Date Rented	Date Returned	Add New Field
1	101	8/30/2008	12/5/2008	
2	102	8/15/2008		
3	103	8/20/2008		
4	104	1/5/2008		
5	105	1/30/2008		
6	106	11/12/2008	12/12/2008	
7	107	9/15/2008		
8	108	10/1/2008		
9	109	8/1/2008	11/30/2008	
*				

FIGURE 1-3 The RENTALS table

Rental Line Item			
Rental ID	Item ID	Quantity	Add New Field
1	D-101	1	
1	D-104	1	
2	C-101	1	
2	D-104	1	
3	S-101	2	
4	L-101	2	
5	C-102	4	
6	D-101	1	
6	S-103	1	
7	C-102	8	
8	L-102	2	
8	L-103	2	
9	S-102	2	
*			

FIGURE 1-4 The RENTAL LINE ITEM table

ASSIGNMENT 2: CREATING A FORM, QUERIES, AND A REPORT

Assignment 2A: Creating a Form

Create a form for easy recording of each rental by FRI. Using the RENTALS table, use the Form Wizard to create this main form. Within this form, create a subform using the RENTAL LINE ITEM table. Save the main form under the filename Rentals. Save the subform as Rental Line Item. View one record and print it if your instructor requires you to print a record. Your output should resemble that in Figure 1-5.

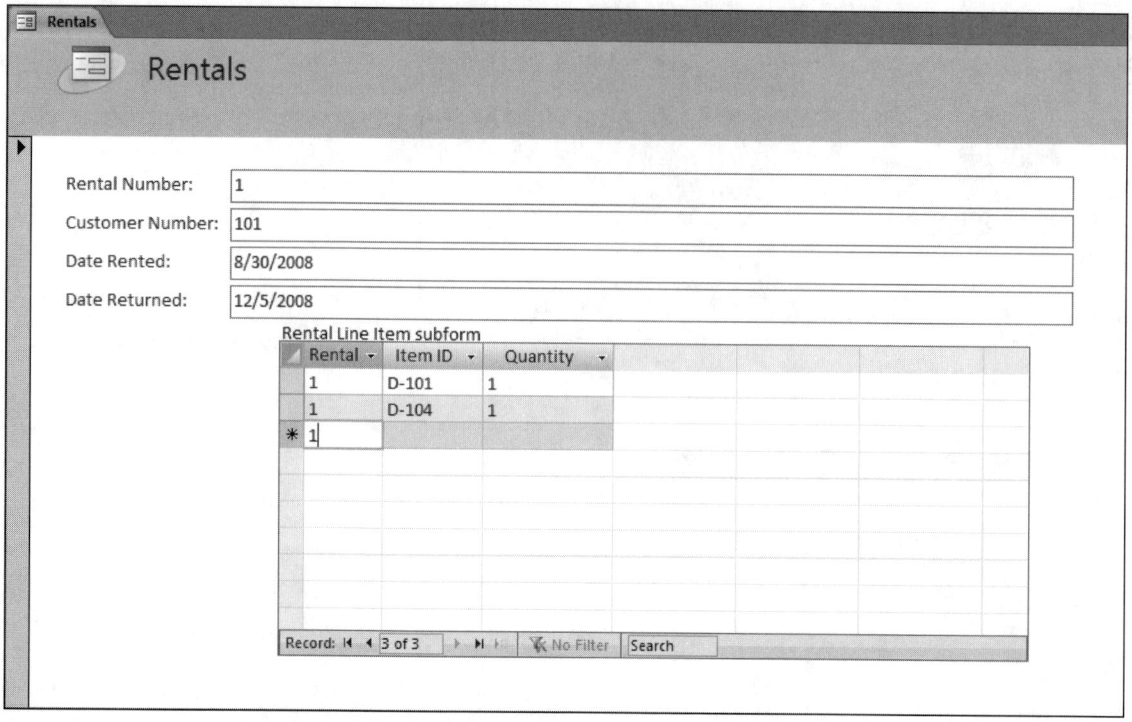

FIGURE 1-5 The Rentals form with subform

Assignment 2B: Creating a Select Query

Create a Select query that lists contact information for customers who have rented defective kitchen chairs. In the output, show Last Name, First Name, Email, Item ID, and Description. Save the query as Defective Item. Your output should resemble that in Figure 1-6.

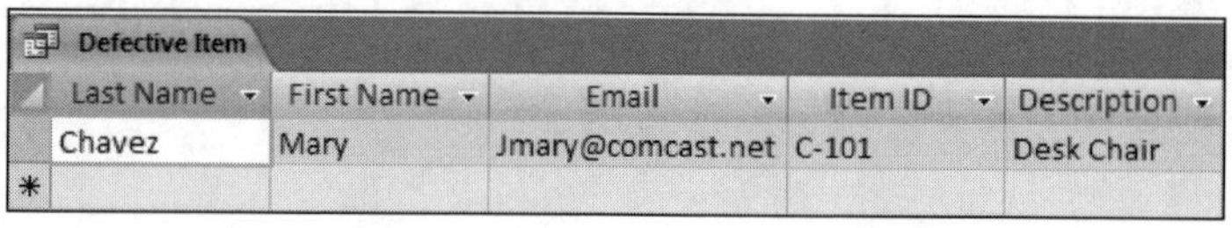

| Defective Item | | | | |
Last Name	First Name	Email	Item ID	Description
Chavez	Mary	Jmary@comcast.net	C-101	Desk Chair
*				

FIGURE 1-6 Defective Item query

Run the query. Print the results, if required.

Assignment 2C: Creating a Select Query

Create a query that lists the renters who have not returned their rental items. In your output, show the Last Name, First Name, Email, and Date Rented. (Hint: Use the criterion of Is Null.) Save your query as Outstanding Rentals. Your output should resemble that in Figure 1-7. Print the output, if desired.

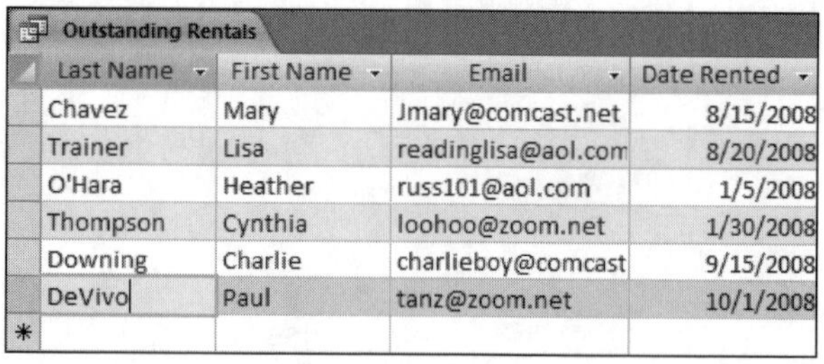

| Outstanding Rentals | | | |
Last Name	First Name	Email	Date Rented
Chavez	Mary	Jmary@comcast.net	8/15/2008
Trainer	Lisa	readinglisa@aol.com	8/20/2008
O'Hara	Heather	russ101@aol.com	1/5/2008
Thompson	Cynthia	loohoo@zoom.net	1/30/2008
Downing	Charlie	charlieboy@comcast	9/15/2008
DeVivo	Paul	tanz@zoom.net	10/1/2008
*			

FIGURE 1-7 Outstanding Rentals query

Assignment 2D: Creating a Totals Query

Create a Totals query that lists the number of items rented, from the most popular to the least popular. Show the Description and Number Rented. Note that the column heading must be changed. Save the query as Popular Items. Your output should resemble that in Figure 1-8. Print the output, if desired.

Popular Items	
Description	Number Rented
Kitchen Chair	12
Triple Sofa	2
Table Lamp	2
Single Dresser	2
Pole Lamp	2
Double Sofa	2
Desk Lamp	2
Desk	2
Sofa Bed	1
Desk Chair	1

FIGURE 1-8   Popular Items query

Assignment 2E: Creating a Query with a Calculated Field

Create a query that calculates the amount of money owed by each renter. In the rent calculation, you can assume 31 days per month and you can use partial months. Make sure you format your output. Your output should include the Customer ID, Last Name, First Name, Email, Description, Date Returned, and Total Bill; it should resemble the output in Figure 1-9. Save the query as Total Bill. Print the output, if desired.

Customer ID	Last Name	First Name	Email	Description	Date Returned	Total Bill
101	Smith	John	Jsmith@pcp.net	Single Dresser	12/5/2008	$93.87
101	Smith	John	Jsmith@pcp.net	Desk	12/5/2008	$87.61
106	Isaacs	Irving	irv@aol.com	Single Dresser	12/12/2008	$29.03
106	Isaacs	Irving	irv@aol.com	Sofa Bed	12/12/2008	$33.87
109	Bevans	Beverly	bev@comcast.net	Triple Sofa	11/30/2008	$117.10

FIGURE 1-9   Total Bill query

Assignment 2F: Generating a Report

Generate a report based on the query in Assignment 2E. To create the report, you need to do the following:

- Using the Report Wizard, base your report on the Total Bill query.
- Group on the Customer ID. Give the report a title of Total Bill.
- Adjust any other parts of the report to resemble the report in Figure 1-10, including moving fields to the Customer ID header.
- Use Print Preview before printing to make sure the report looks correct.

If you are working with a disk or USB memory stick, make sure you do not remove it until after you close the database file.

Total Bill

Customer ID	Name		Email	Description	Total Bill
101	Smith	John	Jsmith@pcp.net		
				Desk	$87.61
				Single Dresser	$93.87
Total for Customer					$181.48
106	Isaacs	Irving	irv@aol.com		
				Sofa Bed	$33.87
				Single Dresser	$29.03
Total for Customer					$62.90
109	Bevans	Beverly	bev@comcast.net		
				Triple Sofa	$117.10
Total for Customer					$117.10
Grand Total					**$361.48**

FIGURE 1-10 Total Bill report

DELIVERABLES

Assemble the following deliverables for your instructor, either electronically or in printed form:

1. Four tables
2. Form: Rentals
3. Query 1: Defective Item
4. Query 2: Outstanding Rentals
5. Query 3: Popular Items
6. Query 4: Total Bill
7. Report: Total Bill
8. Any other required tutorial printouts or electronic media

Staple all pages together. Put your name and class number at the top of the page. If required, make sure your electronic media is labeled.

CASE **2**

VOLUNTEER VACATIONS

Designing a Relational Database to Create Tables,
Forms, Queries, and Reports

PREVIEW

In this case, you'll design a relational database for a business that organizes volunteer trips for college students. After your design is completed and correct, you will create database tables and populate them with data. Then you will produce one form, three queries, and one report. The queries will address the following questions: What trips are available in August? Which students are going to which destination? What are the most popular destinations? Your report will display the amount of money owed for each trip by each student.

PREPARATION

- Before attempting this case, you should have some experience in database design and in using Microsoft Access.
- Complete any part of Database Design Tutorial A that your instructor assigns.
- Complete any part of Access Tutorial B that your instructor assigns, or refer to the tutorial as necessary.
- Refer to Tutorial F as necessary.

BACKGROUND

College students often like to volunteer to help others. They also enjoy traveling to interesting destinations. Those two factors inspired the owners of Volunteer Vacations to create a company that combined both. Volunteer Vacations organizes vacations for college students to destinations in the United States and across the world. Each vacation has a purpose to help people, animals, or nature. Students can choose from trips to help out at camps in the United States for disabled or low-income children, or they can take trips to Sri Lanka to build homes for orphans. If a student likes nature, he or she can work on nature conservation in New Zealand or work to protect the golden lion tamarin, a monkey in Brazil. Housing is provided for all trips and meals are provided on some. The total cost of the trip does not include the flights.

Volunteer Vacations started as a small company and is growing rapidly. You have been hired to create a database to keep track of the trips each summer, all the students who sign up for the trips, and what trips they sign up to take.

You have several goals that you want to accomplish with the database. First, you need to keep track of all the students—their student ID number, address, telephone number, and e-mail address. Of course, you need to keep track of all the trips offered this summer. You need trip information such as the destination, volunteer work, and cost. You also need to note whether the trip includes meals. Keep in mind that some trips go to the same destination. For example, Rio de Janeiro, Brazil, is a destination both for volunteer work for at-risk children and for golden lion tamarin conservation. Finally, you must keep track of which students sign up for which trip. When the students sign up, they must pay an initial deposit. This deposit information must be recorded as well.

As students register for the trips and pay their deposits, the owners of Volunteer Vacations want to be able to record this information directly into the database. You need to create a form to record this data. Eventually, this information will be available on the Web so that students can register online and pay their deposit with a credit card.

Students often call the office and ask which trips are available at certain times during the summer. For example, students who take a session at their university at the beginning of the summer might want to know which trips are available in August this year. You need to create a query to answer that question.

You also need to be able to list students who are going to particular destinations for visa purposes and to track other information. You can create a query that prompts for a destination and then lists all the students who are going to that destination. For marketing purposes, you need to know the most popular destination as well. You can accomplish this task in a query that lists each destination and the number of students going to each destination, from the most popular to the least popular.

Finally, you need to create a report that lists the students' initial payments and their remaining payments required for each trip. This report will ensure that proper bills are sent to each student prior to travel.

ASSIGNMENT 1: CREATING THE DATABASE DESIGN

In this assignment, you will design your database tables on paper using a word-processing program. Pay close attention to the tables' logic and structure. Do not start your Access code (Assignment 2) before getting feedback from your instructor on Assignment 1. Keep in mind that you need to examine the Assignment 2 requirements to design your fields and tables properly. It's good programming practice to look at the required outputs before designing your database. When designing the database, observe the following guidelines:

- First, determine the tables you'll need by listing on paper the name of each table and the fields it should contain. Avoid data redundancy. Do not create a field if it could be created by a "calculated field" in a query.
- Include a logical field that answers the question "Are meals included?"
- You'll need transaction tables. Think about what business events are occurring with each student's actions. Avoid duplicating data.
- Document your tables by using the Table facility of your word processor. Your word-processed tables should resemble the format of the table in Figure 2-1.
- You must mark the appropriate key field(s). You can designate a key field by using an asterisk (*) next to the field name. Keep in mind that some tables need a compound primary key to uniquely identify a record within a table.
- Print the database design.

Table Name	
Field Name	Data Type (such as text, numeric, or currency)
...	...
...	...

FIGURE 2-1   Table design

NOTE

Have your design approved before beginning Assignment 2; otherwise, you may need to redo Assignment 2.

ASSIGNMENT 2: CREATING THE DATABASE AND MAKING QUERIES AND A REPORT

In this assignment, you will first create database tables in Access and populate them with data. Next, you will create a form, three queries, and a report.

Assignment 2A: Creating Tables in Access

In this part of the assignment, you will create your tables in Access. Use the following guidelines:

- Type records into the tables, using the students' names and addresses shown in Figure 2-2. Add your name, address, and e-mail address as an additional student. Make up student IDs, telephone numbers, and e-mail addresses.
- Make up seven trips to various destinations. Use the same destination for two different programs.
- Each student should register for at least one trip. Make most of them register for two. Make up various deposit amounts for each registration.
- Appropriately limit the size of the text fields; for example, a telephone number does not need to have the default setting of 255 characters in length.
- Print all tables.

Students					
Last Name	First Name	Address	City	State	Zip
Burham	Luke	212 Wedgewood Rd	Kennett Square	PA	19348
Doorey	Leon	9 Sandelwood Dr	Avondale	PA	19311
Mattern	Luis	205 Hanover Pl	Avondale	PA	19311
Seals	Lauren	163 Darling Rd	Kennett Square	PA	19348
Webster	Sally	15 Anglin Dr	Wilmington	DE	19808
Ward	Harry	11 Chapel Rd	Wilmington	DE	19808
Seever	Patti	19 Danvers Circle	Kennett Square	PA	19348
Poplawski	Meredith	281 Beverly Rd	Elkton	MD	21119
Meartz	Maria	511 Sparrow Ct	Wilmington	DE	19808
Lee	Gregory	139 Boyer Way	Avondale	PA	19311

FIGURE 2-2 Data

Assignment 2B: Creating Forms, Queries, and a Report

You must generate one form, three queries, and one report, as outlined in the Background section for this case.

Form

Create a form based on your registration table. Save the form as REGISTRATION AND DEPOSIT. Your form should resemble the one shown in Figure 2-3.

FIGURE 2-3 Registration and Deposit form

Query 1

Create a query called August Trips. The output of the query should list the Trip ID, Destination, Purpose, Price, and Start Date for trips beginning in August. Your output should resemble that in Figure 2-4, although your data will be different.

Trip ID	Destination	Purpose	Price	Start Date
104	Auckland, New Zealand	Nature conservation	$900.00	8/5/2008
107	Chicago	Camp Hope - Low income kids	$250.00	8/5/2008

FIGURE 2-4   August Trips query

Query 2

Create a parameter query called Students in Which City. This query needs to prompt the user for a destination. List the Last Name, First Name, and Telephone of the student registered for the destination. If New Orleans was input, your output should resemble that in Figure 2-5, although your data will be different.

Last Name	First Name	Telephone
Seals	Lauren	(610)369-9550
Seever	Patti	(610)544-0923
Lee	Gregory	(610)655-0936

FIGURE 2-5   Students in Which City query

Query 3

Create a query called Most Popular Destination. In this query, you need to add up all the students registered for each destination and display that number with the destination. Show the most popular destination to the least popular destination. Note the column heading change.

Your output should resemble the format in Figure 2-6, but the data will be different.

Destination	Number of Students
Rio de Janeiro, Brazil	5
Chicago	4
New Orleans	3
New Hampshire	1
Colombo, Sri Lanka	1
Auckland, New Zealand	1

FIGURE 2-6   Most Popular Destination query

Report

Create a report called Remaining Payments Required. Your report's output should show headers for Last Name, Destination, Price, Deposit Paid, and Remaining Money Required. Use the following procedure:

1. First create a query for input to the report. The query should contain a calculated field.
2. Bring the query into the Report Wizard.
3. Group on the Last Name field.
4. Sum the Price, Deposit Paid, and Remaining Money Required for each group.
5. Depending on your data, your output should resemble that in Figure 2-7.

Case 2

Remaining Payments Required

Last Name	Destination	Price	Deposit Paid	Remaining Money Required
Burham				
	Rio de Janeiro, Brazil	$2,000.00	$100.00	$1,900.00
	Chicago	$250.00	$50.00	$200.00
Sum		$2,250.00	$150.00	$2,100.00
Doorey				
	Rio de Janeiro, Brazil	$2,000.00	$150.00	$1,850.00
	Chicago	$250.00	$50.00	$200.00
Sum		$2,250.00	$200.00	$2,050.00
Lee				
	New Orleans	$500.00	$50.00	$450.00
	Chicago	$250.00	$25.00	$225.00
Sum		$750.00	$75.00	$675.00
Mattern				
	Colombo, Sri Lanka	$1,100.00	$100.00	$1,000.00
	Chicago	$250.00	$45.00	$205.00
Sum		$1,350.00	$145.00	$1,205.00
Meartz				
	Auckland, New Zealand	$900.00	$400.00	$500.00
Sum		$900.00	$400.00	$500.00
Poplawski				

FIGURE 2-7   Remaining Payments Required report

ASSIGNMENT 3: MAKING A PRESENTATION

Create a presentation that explains the database to the owners of Volunteer Vacations. Include the design of your database tables and how to use the database. Your presentation should take less than 10 minutes, including a brief question-and-answer period.

DELIVERABLES

Assemble the following deliverables for your instructor, either electronically or in printed form:

1. Word-processed design of tables
2. Tables created in Access
3. Form: Registration and Deposit
4. Query 1: August Trips
5. Query 2: Students in Which City
6. Query 3: Most Popular Destination
7. Report: Remaining Payments Required
8. Presentation materials
9. Any other required tutorial printouts or tutorial diskette, CD, or USB memory stick

Staple all pages together. Put your name and class number at the top of the page. Make sure your diskette, CD, or USB memory stick is labeled, if required.

CASE **3**

ONLINE WORKOUTS DATABASE

Designing a Relational Database to Create Tables,
Forms, Queries, Reports, and a Navigation Pane

PREVIEW

In this case, you'll design a relational database for a business that offers personal training over the Internet. After your design is completed and correct, you will create database tables and populate them with data. Then you will produce one form with a subform, four queries, and one report. The form will include customer information, including whether a personal trainer has been assigned and which level of training the customer wants. The queries will address the following questions and tasks: Which customers are over 40 years old? Which customers are very fit or moderately fit and enrolled in the Gold fitness package? How many clients does each trainer have? All members must be over 18 years old, so you will delete records for anyone younger than 18. Your report will summarize the monthly fees brought in by each trainer for each client. You will create a customized navigation pane to manage the database easily.

PREPARATION

- Before attempting this case, you should have some experience in database design and in using Microsoft Access.
- Complete any part of Database Design Tutorial A that your instructor assigns.
- Complete any part of Access Tutorial B that your instructor assigns, or refer to the tutorial as necessary.
- Refer to Tutorial F as necessary.

BACKGROUND

Your best friend, Sally, is a health and exercise science major at your university. The two of you have decided to start a business that offers personal training to customers via the Internet. Here's how it works: Customers register for monthly exercise regimes at three different levels: Gold, Silver, or Bronze. The customers pay a monthly fee for licensed personal trainers to send them a series of exercise workouts, along with links to videos and nutrition information. Sally is in charge of hiring the personal trainers and assembling the various videos and other equipment and material. Because you are proficient in Microsoft Access, you will be in charge of creating the database, which is a prototype system. The final version will be migrated to the Web.

Sally's idea of the business includes recording specific information about the customers. You will not only have to record their information, such as name, address, phone, and e-mail address, you will also have to ask some probing questions to collect data such as birth date, exercise history, fitness level, and medical history. Once the information is recorded, the customer can be assigned a personal trainer and register for a level of training. The Gold level is the most expensive, at $30 per month. That price includes the exercise regime, unlimited e-mails, and customized streaming video from the customer's personal trainer. The next level, Silver at $20, also includes the exercise regime and e-mailing, but it only provides access to prerecorded videos on the Web. Finally, the third level, Bronze at $18.50, includes only the exercise regime and 10 e-mail consultations. Along with this information, the database also needs to keep track of information about personal trainers hired by Online Workouts.

Because this database will eventually be moved to the Web, Sally thinks it's a good idea to create a form with a subform as a prototype of how customers might register for this service. The main form would include all customer information and the subform would include the customer's personal trainer and desired package level (Gold, Silver, or Bronze).

You and Sally have put your heads together and come up with a number of useful queries that should be added to the database. First, it would be good for the personal trainers to be alerted to customers who are over 40 years old. A query could list those clients easily. Another possible question is which customers are very fit or moderately fit in the Gold program. This list could be useful for marketing further products such as a personal nutritionist or exercise clothing. You would also like to keep tabs on how many clients each personal trainer handles. In addition, although you have stated on your Web site that customers must be at least 18 to enroll, some underage people might have signed up. By local law, no one under 18 can be directed remotely by a personal trainer. You can use a query to delete underage customers.

Finally, you and Sally would like to see how much money you are making each month per trainer, along with customer details. This would fit nicely into a monthly report. In addition, you propose creating a custom navigation pane so that users who are unfamiliar with Microsoft Access can easily use the database.

ASSIGNMENT 1: CREATING THE DATABASE DESIGN

In this assignment, you will design your database tables on paper, using a word-processing program. Pay close attention to the tables' logic and structure. Do not start your Access code (Assignment 2) before getting feedback from your instructor on Assignment 1. Keep in mind that you will need to look at the requirements in Assignment 2 to design your fields and tables properly. It's good programming practice to look at the required outputs before designing your database. When designing the database, observe the following guidelines:

- First, determine the tables you need by listing on paper the name of each table and the fields it should contain. Avoid data redundancy. Do not create a field if it could be created by a "calculated field" in a query.
- You'll need a transaction table for recording which trainer is with which customer and the level of program. Avoid duplicating data.
- Document your tables by using the Table facility of your word processor. Your word-processed tables should resemble the format of the table in Figure 3-1.
- You must mark the appropriate key field(s). You can designate a key field by using an asterisk (*) next to the field name. Keep in mind that some tables need a compound primary key to uniquely identify a record within a table.
- Print the database design.

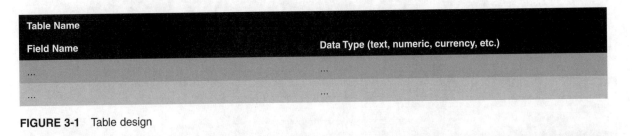

Table Name	
Field Name	Data Type (text, numeric, currency, etc.)
...	...
...	...

FIGURE 3-1 Table design

NOTE

Have your design approved before beginning Assignment 2; otherwise, you may need to redo Assignment 2.

ASSIGNMENT 2: CREATING THE DATABASE, QUERIES, AND A REPORT

In this assignment, you will create database tables in Access and then populate them with data. Next, you will create four queries and a report.

Assignment 2A: Creating Tables in Access

In this part of the assignment, you will create your tables in Access. Use the following guidelines:

- Type records into the tables, using your classmates' and friends' names and addresses. Add your name and address as an additional customer; make sure you have at least 10 customers. Make up customer IDs, telephone numbers, and e-mail addresses. Also, make up birth dates, exercise history, fitness levels, and medical history.
- For the last three fields mentioned in the previous bullet point, use a lookup column for the field choices. You can find the Lookup Column button in the Table Tools Datasheet tab. Use Frequent, Moderate, or None for exercise history; Very fit, moderately fit, or unfit for fitness level; and Very healthy, somewhat healthy, and unhealthy for the medical history choices.
- Assume there are only three personal trainers. Use the names of famous people and make up their addresses, telephone numbers, and e-mail addresses.
- Assume there are three levels of programs and pricing: Gold at $30, Silver at $20, and Bronze at $18.50.
- Register each customer for one trainer and one program. Use the lookup column to look up the three choices.
- Appropriately limit the size of the text fields; for example, a zip code does not need the default setting of 255 characters.
- Print all tables if your instructor requires it.

Assignment 2B: Creating Forms, Queries, and a Report

You need to generate one form with a subform, four queries, one report, and one navigation pane, as outlined in the Background section of this case.

Form

Create a form and subform based on the process of filling out all the customer information and then being assigned a personal trainer and registering for a class level. Save the form as Customer. Print any record from the form, if required. Your form should resemble that in Figure 3-2.

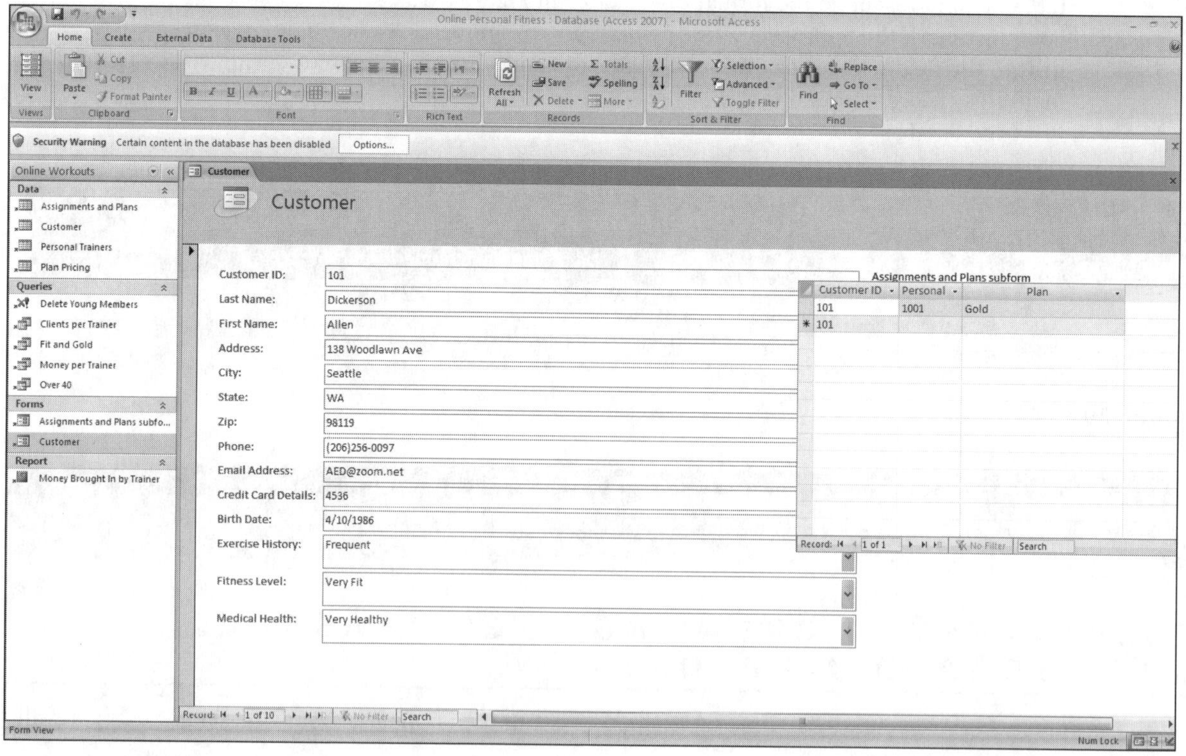

FIGURE 3-2 Customer form

Query 1

Create a query called Over 40. The output of the query should list only the Customer ID, Last Name, First Name, Email Address, and Birth Date. Your output should resemble that shown in Figure 3-3, with different data.

Over 40				
Customer ID	Last Name	First Name	Email Address	Birth Date
102	Faber	Dale	Oldie@brandywine.net	7/12/1958
104	Lavelle	Shirley	Shirl121@hotmail.com	9/25/1960
105	Nelson	Janice	NelsJ@hotmail.com	12/25/1948
109	Trapp	John	Trapp334@hotmail.com	11/26/1967
110	Wills	Billy	BillyW@aol.com	7/1/1951

FIGURE 3-3   Over 40 query

Query 2

Create a query called Fit and Gold. Show Customer ID, Last Name, First Name, and Email Address. List only customers who are very fit or moderately fit and who have joined the Gold program. Your output should look like that in Figure 3-4, although your data will be different.

Fit and Gold			
Customer ID	Last Name	First Name	Email Address
101	Dickerson	Allen	AED@zoom.net
102	Faber	Dale	Oldie@brandywine.net
106	Schwartz	Byron	BS599@aol.com

FIGURE 3-4   Fit and Gold query

Query 3

Create a query called Clients per Trainer. Use the query to add up all the customers for each personal trainer. Display the personal trainer's (PT) Last Name and First Name along with the Number of Clients. Note the column heading change. Your data will differ but your output should resemble that in Figure 3-5.

Clients per Trainer		
PT Last Name	PT First Name	Number of Clients
Gonzalez	Diego	3
Kim	Jon	4
McCall	Patti	3

FIGURE 3-5   Clients per Trainer query

Query 4

Create a query that deletes any record for any customer under the age of 18. Save the query as Delete Young Members.

Report

Create a report called Monthly Money Brought In by Trainer. Your report's output should show headers for Personal Trainer Last Name and First Name, the customer's Last Name and First Name, the Level, and Price Per Month. Use the following procedure:

1. First create a query for input to the report.
2. Group the report on the Personal Trainer Last Name.
3. Sum the Price field.
4. Name the report Monthly Money Brought In by Trainer.
5. Adjust the design view to resemble that in Figure 3-6.

Online Workouts Database

Monthly Money Brought In by Trainer

PT Last Name	PT First Name	Last Name	First Name	Level	Price Per Month
Gonzalez					
	Diego	Sunzar	Sam	Gold	$30.00
	Diego	Schwartz	Byron	Gold	$30.00
	Diego	Dickerson	Allen	Gold	$30.00
Sum					$90.00
Kim					
	Jon	Wills	Billy	Silver	$20.00
	Jon	Nelson	Janice	Silver	$20.00
	Jon	Lavelle	Shirley	Bronze	$18.50
	Jon	Hearn	Arthur	Silver	$20.00
Sum					$78.50
McCall					
	Patti	Trapp	John	Bronze	$18.50
	Patti	Turner	Cynthia	Bronze	$18.50
	Patti	Faber	Dale	Gold	$30.00
Sum					$67.00
Grand Total					$235.50

FIGURE 3-6 Monthly Money Brought In by Trainer report

Navigation Pane

Create a custom navigation pane for easy access to the database. Include the form and the report. Name the navigation pane Online Workouts. Its interface should resemble that in Figure 3-7. Note that the tables are hidden beneath the data to avoid revealing the solutions to Assignment 1.

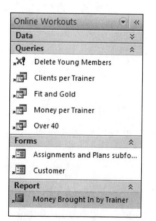

FIGURE 3-7 Online Workouts navigation pane

ASSIGNMENT 3: MAKING A PRESENTATION

Create a presentation that explains the database to Sally and the personal trainers. Include the design of your database tables and how to use the database. Your presentation should take less than 10 minutes, including a brief question-and-answer period.

DELIVERABLES

Assemble the following deliverables for your instructor, either electronically or in printed form:

1. Word-processed design of tables
2. Tables created in Access
3. Form: Customer (one record printed, if required)
4. Query 1: Over 40
5. Query 2: Fit and Gold
6. Query 3: Clients per Trainer
7. Query 4: Delete Young Members
8. Report: Monthly Money Brought In by Trainer
9. Navigation pane: Online Workouts
10. Presentation materials
11. Any other required tutorial printouts or tutorial diskette, CD, or USB memory stick

Staple all pages together. Put your name and class number at the top of the page. Make sure your diskette, CD, or USB memory stick is labeled, if required.

CASE **4**

SILENT CAR AUCTION DATABASE

Designing a Relational Database to Create Tables, Forms, Queries, Reports, and Navigation Panes

PREVIEW

In this case, you'll design a relational database for a monthly silent car auction by a local bank. After your database design is completed and correct, you will create database tables and populate them with data. Then you will produce a form, five queries, a report, and a custom navigation pane. The form will record all bids for automobiles. The queries will display all newer automobiles that are available, all Chevrolets, and all Fords. Other queries will report the maximum bid and the winners and losers of each auction. The reports will summarize the available automobiles by year of manufacture. The custom navigation pane will allow access to all tables, forms, queries, and the report.

PREPARATION

- Before attempting this case, you should have some experience in database design and in using Microsoft Access.
- Complete any part of Database Design Tutorial A that your instructor assigns.
- Complete any part of Access Tutorial B that your instructor assigns, or refer to the tutorial as necessary.
- Refer to Tutorial F as necessary.

BACKGROUND

A local bank, Merchant Savings Society, offers automobile loans to many types of customers, including some with shady credit histories. Subsequently, Merchant ends up repossessing an average of 25 cars per month. Once the cars are repossessed, Merchant sells them to individual buyers to recoup some of their losses. The cars are placed in a parking lot owned by Merchant; each car has the keys in the ignition. Potential buyers are invited to inspect the cars and start them, but they cannot drive the cars. To find out which cars are available each month, potential buyers subscribe to a mailing list. Buyers can look over the list of cars and decide if they want to inspect one. If the car looks good, the buyer puts in a bid to Merchant via e-mail. The highest bid wins the car. Merchant has been doing the paperwork for these transactions by hand, but management feels it's time to computerize the system because it's quite popular and many potential customers now participate. Because you are proficient in Microsoft Access, you have been hired to create a database for Merchant.

You must consider a number of parameters when designing the database for Merchant. When the bank repossesses a car and places it in the lot, bank employees note the location in the lot so they can find it easily. They also write down the year, type of car, mileage, and the VIN (vehicle identification number). Potential customers' information is also kept in an Excel spreadsheet. Their name and address are recorded along with e-mail addresses. Finally, the bank needs to have some way of recording the bids that come in. Customers send in their bids on specific cars. The bids are recorded and the date is noted. A form would be a handy way to record these bids for each car.

Frequently, customers like to look at particular manufacturers' cars that are available. The bank would like to run some queries each month to list available Chevrolet and Ford cars. Also, customers are mostly interested in cars that are only a few years old, because they are the most economical. So, the bank would like a way of listing only cars that are two or three years old.

Of course, the bank needs to know the top bid for each auction, and would like to have a query that shows the top bid for each vehicle. Bank employees also would like to enter an Auto ID and see which customers won their bids and which lost for a specific car.

For advertising purposes, the bank would like a report of all available cars, grouped by year of manufacture. Finally, the bank would like a custom navigation pane to make it easy to work with the database.

ASSIGNMENT 1: CREATING THE DATABASE DESIGN

In this assignment, you will design your database tables on paper, using a word-processing program. Pay close attention to the tables' logic and structure. Do not start your Access code (Assignment 2) before getting feedback from your instructor on Assignment 1. Keep in mind that you will need to look at the requirements in Assignment 2 to design your fields and tables properly. It's good programming practice to look at the required outputs before designing your database. When designing the database, observe the following guidelines:

- First, determine the tables you need by listing on paper the name of each table and the fields it should contain. Avoid data redundancy. Do not create a field if it could be created by a "calculated field" in a query.
- You'll need a transaction table. Avoid duplicating data.
- Document your tables by using the Table facility of your word processor. Your word-processed tables should resemble the format of the table in Figure 4-1.
- You must mark the appropriate key field(s). You can designate a key field by using an asterisk (*) next to the field name. Keep in mind that some tables might need a compound primary key to uniquely identify a record within a table.
- Print the database design.

Table Name	
Field Name	Data Type (text, numeric, currency, etc.)
...	...
...	...

FIGURE 4-1 Table design

NOTE

Have your design approved before beginning Assignment 2; otherwise, you may need to redo Assignment 2.

ASSIGNMENT 2: CREATING THE DATABASE AND MAKING QUERIES AND A REPORT

In this assignment, you will create database tables in Access and then populate them with data. Next, you will create a form, five queries, a report, and a custom navigation pane.

Assignment 2A: Creating Tables in Access

In this part of the assignment, you will create your tables in Access. Use the following guidelines:

- Type records into the tables, using the names and addresses of your friends. Create at least 10 customer records.
- Assume that 25 vehicles are available for bidding.

- Make each customer bid on a vehicle. Have some customers bid on the same vehicle. Assume that all this data is for one month.
- Appropriately limit the size of the text fields; for example, a zip code does not need the default setting of 255 characters.
- Print all tables.

Assignment 2B: Creating Forms, Queries, Reports, and Custom Navigation Pane

You need to create one form, five queries, one report, and one custom navigation pane, as outlined in the Background section of this case.

Form

Create a form that displays the information for each available automobile. Within the form, create a subform that allows Merchant to record all the bids for a car. Your data will vary but the output should resemble that in Figure 4-2.

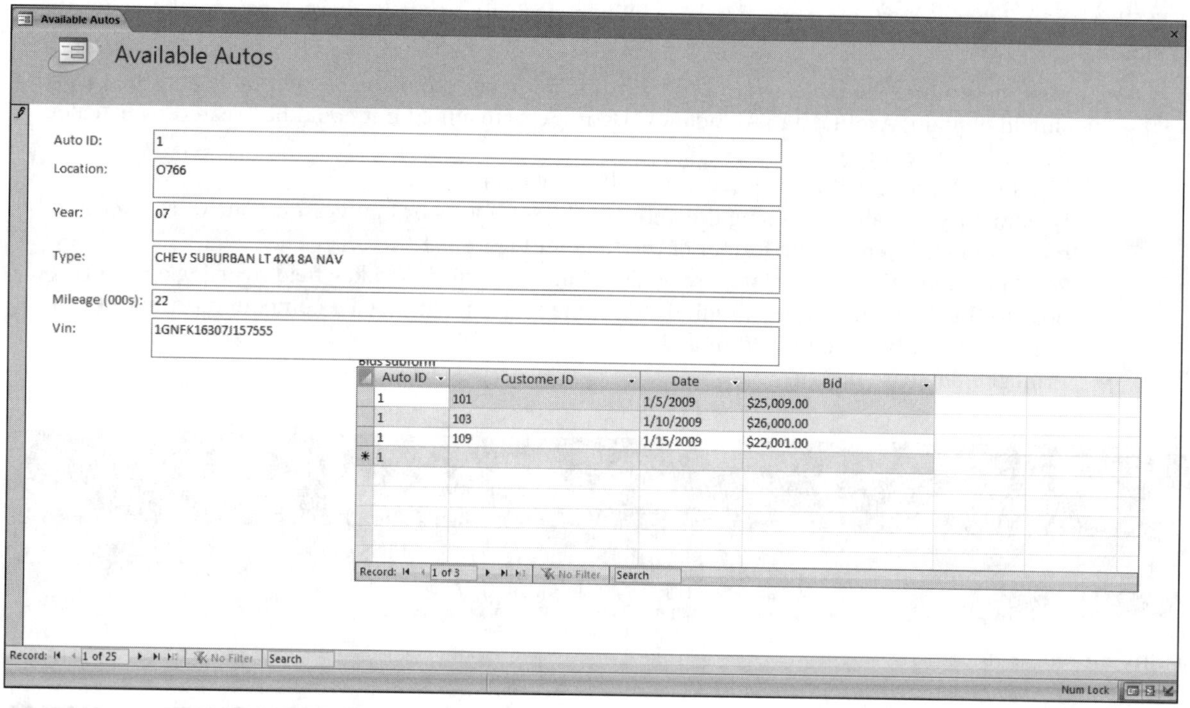

FIGURE 4-2   Available Autos form

Query 1

Create a query called All 2007 and 2006. List the Auto ID, Location, Year, Type, and Mileage for cars manufactured in 2006 and 2007. Depending on the data, the query output might resemble that in Figure 4-3.

Auto ID	Location	Year	Type	Mileage (000s)
1	O765	07	CHEV SUBURBAN LT 4X4 8A NAV	23
2	C173	07	CHEV HHR 4A ASBWLTWCCSR	7
3	L234	06	FORD F150 XL 6A ASBTW	22
4	B155	06	CHEV TAHOE 4X4 8A ASBWLTWCC	48
5	H156	06	PONTIAC G6 4A ASBWLTWCC	19

FIGURE 4-3   All 2007 and 2006 query

Query 2

Create a query called Available Chevrolets. Display the Auto ID, Location, Year, Type, Mileage, and VIN. Your data will differ, but your output should resemble that in Figure 4-4.

Available Chevrolets					
Auto ID	Location	Year	Type	Mileage (000s)	Vin
1	O765	07	CHEV SUBURBAN LT 4X4 8A NAV	23	1GNFK16307J157555
2	C173	07	CHEV HHR 4A ASBWLTWCCSR	7	3GNDA33PX7S5221231
4	B155	06	CHEV TAHOE 4X4 8A ASBWLTWCC	48	1GNEK13T26R105776
13	S126	04	CHEV MALIBU 4A ASBWLTWCC	90	1G1ND52FX4M505982
18	B205	03	CHEV MONTE CARLO SS 6A ASBSR	51	2G1WX12K539362765

FIGURE 4-4   Available Chevrolets query

Query 3

Create a similar query called Available Fords. Your data will differ, but the output should resemble that in Figure 4-5.

Available Fords					
Auto ID	Location	Year	Type	Mileage (000s)	Vin
3	L234	06	FORD F150 XL 6A ASBTW	22	1FTRF12216NB66324
9	C211	05	FORD FOCUS ZX4 4A ASB	45	1FAFP34N45W254872
10	L211	05	FORD F150 XL 4X2 6M ASBTW	49	1FTRF122X5NB81984
12	B196	04	FORD F150 OFFROAD 4X4 XCAB 8	60	1FTPX14504FA15423
23	Z095	02	FORD EXPLORER EB 4X4 8A ASBW	82	1FMDU74W02UB72009
24	D135	02	FORD EXPLORER XLS 4X4 6A ASB	110	1FMZU72E42UA13223
25	F195	02	FORD F150XLT SPORT 4X2 6A A	81	1FTRX17282NA98776

FIGURE 4-5   Available Fords query

Query 4

Create a query called Max bid. This query should display the Auto ID, Location, Year, and Maximum Bid for each auto. Your output should resemble that in Figure 4-6, with different data. Note the column heading change.

Max bid			
Auto ID	Location	Year	Maximum Bid
1	O765	07	$26,000.00
2	C173	07	$14,500.00

FIGURE 4-6   Max(imum) Bid query

Query 5

Create a query called Winners and Losers that prompts the user for an input. The input would be a specific Auto ID. Display the Auto ID, Last Name, Bid, Maximum Bid, and Won Bid?. Hint: Use an If statement and the previous query as input to this query.

Winners and Losers				
Auto ID	Last Name	Bid	Maximum Bid	Won Bid?
1	Burham	$25,009.00	$26,000.00	no
1	Mattern	$26,000.00	$26,000.00	yes
1	Meartz	$22,001.00	$26,000.00	no

FIGURE 4-7   Winners and Losers query

Report

Create a report called Available Autos by Year. Your report's output should show headers for Year, Location, Type, Mileage, VIN, and Auto ID. Group by Year. Your data will differ, but your output should resemble that in Figure 4-8.

Available Autos by Year

Year	Location	Type	Mileage (000s)	Vin	Auto ID
02					
	D135	FORD EXPLORER XLS 4X4 6A ASB	110	1FMZU72E42UA13223	24
	Z095	FORD EXPLORER EB 4X4 8A ASBW	82	1FMDU74W02UB72009	23
	M215	DODGE DAKOTA SLT 4X4 8A ASB	114	1B7HG48NX2S653783	22
	S195	DODGE STRATUS SE 4A ASBWLSR	86	1B3EL46R22N193781	21
	G205	KIA SEDONA LX 6A ASBWLTWCC	53	KNDUP13102633908	20
	F195	FORD F150XLT SPORT 4X2 6A A	81	1FTRX17282NA98776	25
03					
	J225	MAZDA B3000 DUAL SPORT 6M AS	64	4F4YR12U13TM036512	19
	B205	CHEV MONTE CARLO SS 6A ASBSR	51	2G1WX12K539362765	18
	W066	JEEP GR CHKE LAREDO 4X4 6A A	61	1J4GW48S03C591239	17
	H216	KIA SEDONA EX 6A ASBWLTWCCSR	71	KNDUP131236490873	16
04					
	S206	NISSAN SENTRA SE/R 4M ASBSR	41	3N1AB51D24L480863	14
	S126	CHEV MALIBU 4A ASBWLTWCC	90	1G1ND52FX4M505982	13
	H186	LINCOLN NAVIGATOR 8A ASBWLSR	58	5LMFU28R34LJ15871	15
	B196	FORD F150 OFFROAD 4X4 XCAB 8	60	1FTPX14504FA15423	12
05					
	D196	PONTIAC G6 6A ASBWLTWCC	43	1G2ZG528754159894	11

FIGURE 4-8   Available Autos by Year report

Custom Navigation Pane

To access the database contents easily, create a custom navigation pane called Silent Car Auction. Your custom navigation pane should look like that in Figure 4-9. Note that the tables are not shown to avoid revealing the solution to Assignment 1.

FIGURE 4-9   Silent Car Auction custom navigation pane

ASSIGNMENT 3: MAKING A PRESENTATION

Create a presentation for the management of Merchant Savings Society. Pay particular attention to database users who might not be familiar with Microsoft Access. Your presentation should take less than 15 minutes, including a brief question-and-answer period.

DELIVERABLES

Assemble the following deliverables for your instructor, either electronically or in printed form:

1. Word-processed design of tables
2. Tables created in Access
3. Form: Available Autos
4. Query 1: All 2007 and 2006
5. Query 2: Available Chevrolets
6. Query 3: Available Fords
7. Query 4: Max Bid
8. Query 5: Winners and Losers
9. Report: Available Autos by Year
10. Navigation Pane: Silent Car Auction
11. Any other required tutorial printouts or tutorial CD or memory stick

Staple all pages together. Put your name and class number at the top of the page. Make sure your CD or memory stick is labeled, if required.

THE CHARITY DATABASE

Designing a Relational Database to Create Tables, Forms, Queries, Reports, and a Navigation Pane

PREVIEW

In this case, you'll design a relational database for a company that keeps track of charities and their donors. After your database design is completed and correct, you will create database tables and populate them with data. Then you will produce one form and one subform, five queries, two reports, and a custom navigation pane. The form and subform will allow the company to easily record donations to each charity. The queries will allow the management, donors, and charities to access important information. The queries will list charities rated by their program expenses, donors in a specific category of charity, top givers in a specific charity, frequency of giving by donors, and matching gifts for a specific charity. The first report will list the charities by category; the second will show all charitable gifts in a specific time period. The custom navigation pane will allow easy access to all parts of the database.

PREPARATION

- Before attempting this case, you should have some experience in database design and in using Microsoft Access.
- Complete any part of Database Design Tutorial A that your instructor assigns.
- Complete any part of Access Tutorial B that your instructor assigns, or refer to the tutorial as necessary.
- Refer to Tutorial E as necessary.

BACKGROUND

Charitable organizations are big business. Although most people who give to charities see it as a "one-time" donation, charities view a donation as the beginning of a lucrative relationship with the donor. Studies have shown that the three months after a charitable donation are critical in cementing the donor/charity relationship. Donors who give again within three months are more likely to contribute than those who give 12 months after the initial gift. Charities walk a fine line: They want to contact donors for additional gifts, but don't want to pester them. Also, mailings to prospective donors can be very expensive. Charities should use sophisticated statistical analysis to figure out who would be a good donor, like marketers use to figure out what customers will buy.

A college friend of yours, Chuck Ikeda, is interested in charities both from the organizational side and the donors' side. He has started a small company called Charity Info to track charities' statistics and donors' statistics. His vision is to share charity statistics with potential donors and help them to select the best charity for their money and interests. Chuck also will help charities in tracking donors' habits and analyze where they can best put their fundraising resources. He has already obtained a few accounts and realizes that the information must be organized in a database to make it useful to the charity or donor. You have been asked to design and implement a charity database because you know about database design and Microsoft Access.

You must keep several parameters in mind when designing the charity database. For example, you must keep information on the donors. All donors are assigned a unique ID number when they register to donate. Information about names, addresses, and telephone numbers is also recorded. To track charities, you must record information that includes the charity's name, address, and telephone number, along with statistics for program expenses, administrative and fundraising expenses, and recent revenue. Program expenses are the percentage of the charity's total budget that is spent on programs for its benefactors. Likewise, administrative expenses are the percentage of the budget spent on administering the charity and fundraising. If fundraising expenses are reduced by Charity Info's analysis, then more money will be available for needed programs.

Each charity also fits under a specific category, such as animals or environmentalism, that allows potential donors to narrow their search for a charity. Finally, all donations must be recorded.

Chuck would like to have a way to record donations from each donor easily. After donor and charity information is in the database, he wants to be able to enter donation information quickly for repeat donors. You suggest using a form and subform to expedite this chore. Eventually, this form can be migrated to the Web for self-service.

Chuck also has a number of questions for the database. For example, donors call and ask about a charity's program expenses. Often their decision to give to an organization depends on whether most of the money goes to the needy or to administrative costs. Therefore, you need to create a query that lists the charities rated by their program expenses. Then, for marketing purposes, charities want to know who donates to specific categories of charities such as arts, environment, or human services. A query can answer such questions easily.

Within each charity, Chuck wants to know the top donors. In addition, charities often would like to see the frequency of giving by donors and their average gifts, because this affects their soliciting of future gifts. Queries can answer both requests. Finally, for some charities, anonymous donors will match existing donations. This has already happened with a charity in Phoenix, Arizona, that deals with international hunger. An anonymous donor has agreed to match 50 cents for every dollar donated. A query can calculate the donations and total both the original and the matching donation.

Charity Info also has requested some reports. The first report will be displayed on its Web site to show the charities available by category. The second report is a wrap-up of all donations by charity within a specific time period. Finally, a custom navigation pane should be created to organize the use of the database.

ASSIGNMENT 1: CREATING THE DATABASE DESIGN

In this assignment, you will design your database tables on paper using a word-processing program. Pay close attention to the tables' logic and structure. Do not start your Access code (Assignment 2) before getting feedback from your instructor on Assignment 1. Keep in mind that you need to look at the requirements in Assignment 2 to design your fields and tables properly. It's good programming practice to look at the required outputs before designing your database. When designing the database, observe the following guidelines:

- First, determine the tables you'll need by listing on paper the name of each table and the fields it should contain. Avoid data redundancy. Do not create a field if it could be created by a "calculated field" in a query.
- You'll need at least one transaction table. Avoid duplicating data.
- Document your tables by using the Table facility of your word processor. Your word-processed tables should resemble the format of the table in Figure 5-1.
- You must mark the appropriate key field(s). You can designate a key field using an asterisk (*) next to the field name. Keep in mind that some tables might need a compound primary key to uniquely identify a record within a table.
- Print the database design.

Table Name	
Field Name	Data Type (text, numeric, currency, etc.)
...	...
...	...

FIGURE 5-1 Table design

ASSIGNMENT 2: CREATING THE DATABASE AND MAKING QUERIES AND REPORTS

In this assignment, you will first create database tables in Access and populate them with data. Next, you will create a form, five queries, two reports, and a navigation pane.

Assignment 2A: Creating Tables in Access

In this part of the assignment, you will create your tables in Access. Use the following guidelines:

- Type records into the tables, using your classmates and their addresses as the donors. Create at least 11 donors.
- Assume that eight categories of charities exist: animals, environment, international, arts, health, public benefit, education, and human services.
- Choose two of your favorite categories and create three charities for each, for a total of six charities. You can search on the Web for real charities and statistics.
- Have each donor donate once. Have at least five of them donate twice or more—some to the same charity and some to different charities.
- Appropriately limit the size of the text fields; for example, a zip code does not need the default setting of 255 characters.
- Print all tables.

Assignment 2B: Creating Forms, Queries, Reports, and a Custom Navigation Pane

You need to create one form and one subform, five queries, two reports, and one navigation pane, as outlined in the Background section of this case.

Form

Create a form with a subform that allows the employee who answers the telephone to record donations for a specific donor. Your data will vary, but when you view one record, it should resemble Figure 5-2.

Query 1

Create a query called Charities rated by Program Expenses. This query should list the Charity Name, Address, City, State, Zip, and Program Expenses. List the highest Program Expenses for the charity to the lowest. Output to this query should resemble Figure 5-3, although the data will vary.

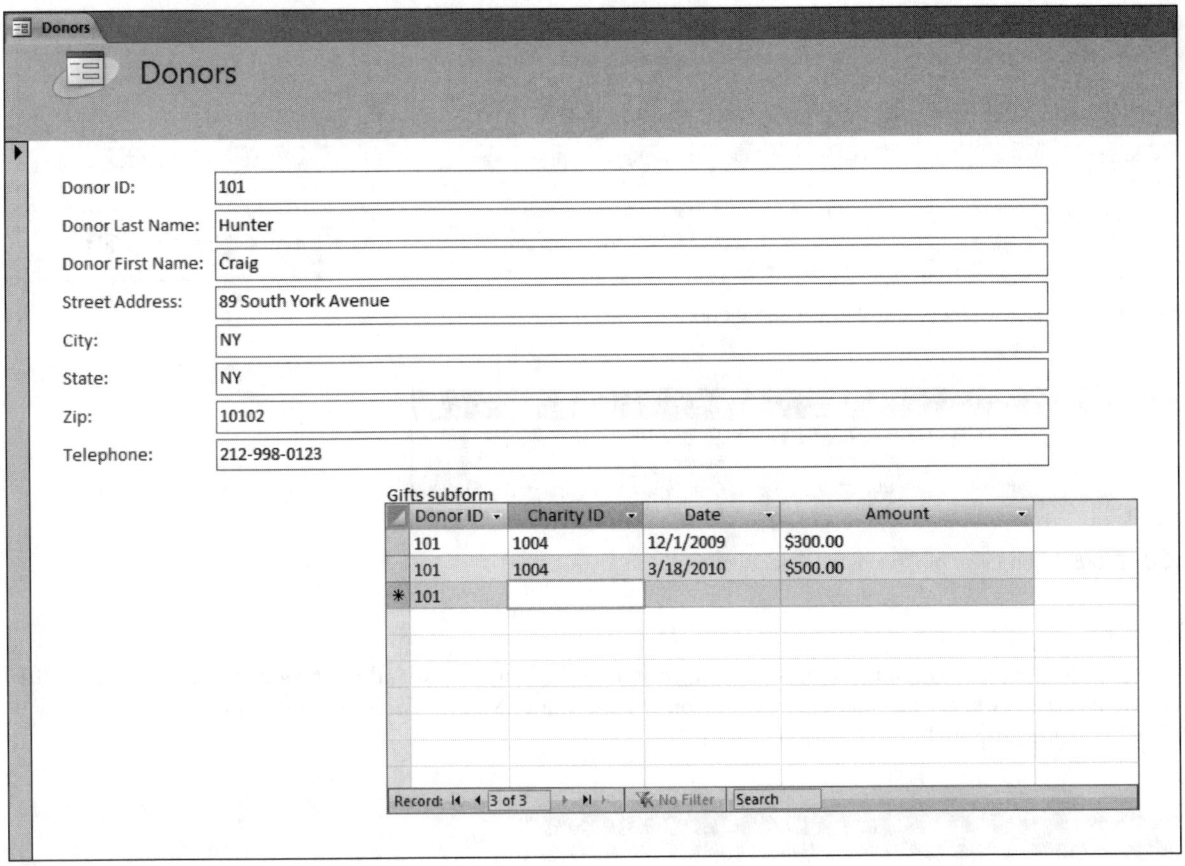

FIGURE 5-2 Donors form

Charities rated by Program Expenses

Charity Name	Charity Address	Charity City	Charity State	Charity Zip	Program Expenses
United Nations Foundation	1225 Connecticut Avenue NW	Washington DC		20036	97
CARE	151 Ellis Street, NE	Atlanta	GA	30303	92
Food for the Hungry	1224 East Washington Street	Phoenix	AZ	85034	89
American Society for the Prevention of Cruelty to Animals	424 East 92nd Street	New York	NY	10128	81
National Aquarium in Baltimore	501 East Pratt Street	Baltimore	MD	21202	72
The Gorilla Foundation	P.O. Box 620530	Woodside	CA	94062	54
*					

FIGURE 5-3 Charities rated by Program Expenses query

Query 2

Create a query called Donors by Requested Category. List the Category, Donor Last and First Names, Street Address, City, State, Zip, and Telephone. Your query should prompt for a category and allow the user to enter the desired category (Animals in this example). Your data will differ, but your output should resemble Figure 5-4.

Query 3

Create a query called Top Giver of Requested Charity. Display the Charity ID, Charity Name, Donor Last and First Names, and Total amount given. This query should prompt for a desired Charity ID. Keep in mind that some donors give more than once, but you want to display the total amount given. Also, your output needs to be sorted from top giver to lowest giver. Note the change to the column heading. Your data will differ, but the output should resemble Figure 5-5.

Donors by Requested Category							
Category	Donor Last Name	Donor First Name	Street Address	City	State	Zip	Telephone
Animals	Cool	DJ	900-A South Chapel St	Newark	DE	19711	302-737-0098
Animals	John	Sanchez	245 South Park Ave	Newark	DE	19711	302-998-6754
Animals	Trent	Barbara	713 Bluefield Road	Chili	NY	14526	585-998-8362
Animals	West	Mary	56 W 103 St	NY	NY	10102	212-998-0987
Animals	Trent	Barbara	713 Bluefield Road	Chili	NY	14526	585-998-8362
Animals	West	Johnny	56 W 103 St	NY	NY	10102	212-998-0987
Animals	West	Johnny	56 W 103 St	NY	NY	10102	212-998-0987

FIGURE 5-4   Donors by Requested Category query

Top Giver of Requested Charity				
Charity ID	Charity Name	Donor Last Name	Donor First Name	Total amount
1004	CARE	Richards	Bernie	$800.00
1004	CARE	Hunter	Craig	$800.00
1004	CARE	Fleisher	Peggy	$15.00

FIGURE 5-5   Top Giver of Requested Charity query

Query 4

Create a query called Frequency of Giving by Donor. List the Donor Last and First Names, the Number of Gifts, and the Average Gift. Sort alphabetically on Donor Last Name. Note the column headings' changes. Data will differ, but your output should look like Figure 5-6.

Frequency of Giving by Donor			
Donor Last Name	Donor First Name	Number of Gifts	Average Gift
Cool	DJ	1	$50.00
Cressman	Anita	1	$65.00
Fleisher	Emily	1	$50.00
Fleisher	Peggy	3	$15.00
Hunter	Craig	2	$400.00
John	Sanchez	1	$1,000.00
Merit	Arthur	1	$55.00
Richards	Bernie	1	$800.00
Trent	Barbara	2	$75.00
West	Johnny	2	$125.00
West	Mary	1	$125.00

FIGURE 5-6   Frequency of Giving by Donor query

Query 5

Create a query called Matching Gifts that displays the Charity Name, Date, and Amount, and then calculate the Matched amount and the Total Donation. Recall that an anonymous donor is willing to match 50 cents on every dollar donated to the Food for the Hungry charity. Data will differ, but your output should resemble Figure 5-7.

Matching Gifts				
Charity Name	Date	Amount	Matched amount	Total Donation
Food for the Hungry	3/16/2010	$15.00	$7.50	$22.50
Food for the Hungry	3/1/2010	$55.00	$27.50	$82.50

FIGURE 5-7   Matching Gifts query

Report 1

Create a report called Charities by Category. Show the Category, Charity Name, Address, and Telephone. You first need to create a query as input to the Report Wizard. Group the report on Category. Your output should resemble Figure 5-8.

Charities by Category

Category	Charity Name	Charity Address	Charity City	Charity State	Charity Zip	Charity Telephone
Animals						
	National Aquarium in Baltimore	501 East Pratt Street	Baltimore	MD	21202	410-576-3800
	The Gorilla Foundation	P.O. Box 620530	Woodside	CA	94062	800-634-6273
	American Society for the Prevention of Cruelty to Animals	424 East 92nd Street	New York	NY	10128	800-628-0028
Environm						
	Food for the Hungry	1224 East Washingto	Phoenix	AZ	85034	800-248-6437
	United Nations Foundation	1225 Connecticut Ave	Washington DC		20036	202-887-9021
	CARE	151 Ellis Street, NE	Atlanta	GA	30303	800-521-2273

Friday, September 21, 2007 — Page 1 of 1

FIGURE 5-8   Charities by Category report

Report 2

Create a report called Charitable Gifts September 2009 – April 2010. Show the Charity Name, Donor Last and First Names, Date, and Amount. You first need to create a query as input to the Report Wizard. Group on Charity Name and sum the Amount. Your output should resemble Figure 5-9. (Note that not all data is visible.)

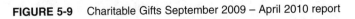

Charitable Gifts September 2009 - April 2010

Charity Name	Donor Last Name	Donor First Name	Date	Amount
American Society for the Prevention of Cruelty to Animals				
	Cool	DJ	2/12/2010	$50.00
	John	Sanchez	10/15/2009	$1,000.00
Total Gift				$1,050.00
CARE				
	Hunter	Craig	3/18/2010	$500.00
	Hunter	Craig	12/1/2009	$300.00
	Fleisher	Peggy	3/1/2010	$15.00
	Richards	Bernie	11/30/2009	$800.00
Total Gift				$1,615.00
Food for the Hungry				
	Merit	Arthur	3/1/2010	$55.00
	Fleisher	Peggy	3/16/2010	$15.00
Total Gift				$70.00
National Aquarium in Baltimore				
	West	Mary	4/14/2010	$125.00
	Trent	Barbara	4/14/2010	$75.00
Total Gift				$200.00
The Gorilla Foundation				

FIGURE 5-9   Charitable Gifts September 2009 – April 2010 report

Custom Navigation Pane

Create a custom navigation pane named Charity Database to access all information in the database easily.

ASSIGNMENT 3: MAKING A PRESENTATION

Create a presentation for Chuck Ikeda and the employees and volunteers at Charity Info. Show the group how to use the database. Include suggestions for further work such as more queries and statistical analysis on the data, and suggestions for more use of the charity data such as its revenue and other expenses. Your presentation should take less than 15 minutes, including a brief question-and-answer period.

DELIVERABLES

Assemble the following deliverables for your instructor, either electronically or in printed form:

1. Word-processed design of tables
2. Tables created in Access
3. Form: Donors
4. Query 1: Charities rated by Program Expenses
5. Query 2: Donors by Requested Category
6. Query 3: Top Giver of Requested Charity
7. Query 4: Frequency of Giving by Donor
8. Query 5: Matching Gifts
9. Report 1: Charities by Category
10. Report 2: Charitable Gifts September 2009 – April 2010
11. Navigation pane: Charity Database
12. Presentation materials
13. Any other required tutorial printouts or tutorial diskette, CD, or USB memory stick

Staple all pages together. Put your name and class number at the top of the page. Make sure your diskette, CD, or USB memory stick is labeled, if required.

PART 2

DECISION SUPPORT CASES
USING EXCEL SCENARIO MANAGER

BUILDING A DECISION SUPPORT SYSTEM IN EXCEL

A **decision support system (DSS)** is a computer program that can represent, either mathematically or symbolically, a problem that a user needs to solve. Such a representation is, in effect, a model of a problem.

Here's how a DSS program works: The DSS program accepts input from the user or looks at data in files on a disk. Then the DSS program runs the input and any other necessary data through the model. The program's output is the information the user needs to solve a problem. Some DSS programs recommend a solution to a problem.

A DSS can be written in any programming language that lets a programmer represent a problem. For example, a DSS can be built in a third-generation language such as Visual Basic or in a database package such as Access. A DSS also can be written in a spreadsheet package such as Excel.

The Excel spreadsheet package has standard built-in arithmetic functions as well as many statistical and financial functions. Thus, many kinds of problems—such as those in accounting, operations, and finance—can be modeled in Excel.

This tutorial has the following four sections:

1. **Spreadsheet and DSS Basics:** In this section, you'll learn how to create a DSS program in Excel. Your program will be in the form of a cash flow model. This section will give you practice in spreadsheet design and in building a DSS program.
2. **Scenario Manager:** In this section, you'll learn how to use an Excel tool called the Scenario Manager. With any DSS package, one problem with playing "what if" is this: Where do you physically record the results from running each set of data? Typically, a user writes the inputs and related results on a sheet of paper. Then—ridiculously enough—the user might have to input the data *back* into a spreadsheet for further analysis. The Scenario Manager solves that problem. It can be set up to capture inputs and results as "scenarios," which are then summarized on a separate sheet in the Excel workbook.
3. **Practice Using Scenario Manager:** You will work on a new problem, a case using the Scenario Manager.
4. **Review of Excel Basics:** This brief section reviews the information you'll need to do the spreadsheet cases that follow this tutorial.

SPREADSHEET AND DSS BASICS

Assume it is late 2008 and you are trying to build a model of what a company's net income (profit) and cash flow will be the next two years (2009 and 2010). This is the problem: to forecast net income and cash flow in those years. The company is likely to use the forecasts to answer a business question or make a strategic decision, so the estimates should be as accurate as you can make them. After researching the problem, you decide that the estimates should be based on three things: (1) 2008 results, (2) estimates of the underlying economy, and (3) the cost of products the company sells.

Your model will use an income statement and cash flow framework. The user can input values for two possible states of the economy in 2009–2010: an *O* for an optimistic outlook or a *P* for a pessimistic outlook. The state of the economy is expected to affect the number of units the company can sell as well as each unit's selling price. In a good, or *O*, economy, more units can be sold at a higher price. The user also can input values into your model for two possible cost-of-goods-sold price directions: a *U* for up or a *D* for down. A *U* means that the cost of an item sold will be higher than it was in 2008; a *D* means that the cost will be less.

Presumably, the company will do better in a good economy and with lower input costs—but how much better? Such relationships are too complex for most people to assess in their head, but a software model can assess them easily. Thus, the user can play "what if" with the input variables and note the effect on net income and year-end cash levels. For example, a user can ask these questions: What if the economy is good and costs go up? What will net income and cash flow be in that case? What will happen if the economy is down and costs go down? What will be the company's net income and cash flow in that case? With an Excel software model available, the answers to those questions are easily quantified.

Organization of the DSS Model

Your spreadsheets should have the following sections, which will be noted in boldfaced type throughout this tutorial and in the Excel cases featured in this textbook:

- **Constants**
- **Inputs**
- **Summary of Key Results**
- **Calculations** (of values that will be used in the **Income Statement and Cash Flow Statement**)
- **Income Statement and Cash Flow Statement**

Here as an extended illustration, a DSS model is built for the forecasting problem described. Next, you'll look at each spreadsheet section. Figure C-1 and Figure C-2 show how to set up the spreadsheet.

	A	B	C	D
1	**TUTORIAL EXERCISE**			
2				
3	**CONSTANTS**	**2008**	**2009**	**2010**
4	TAX RATE	NA	0.33	0.35
5	NUMBER OF BUSINESS DAYS	NA	300	300
6				
7	**INPUTS**	**2008**	**2009**	**2010**
8	ECONOMIC OUTLOOK (O = OPTIMISTIC; P = PESSIMISTIC)	NA		NA
9	PURCHASE-PRICE OUTLOOK (U = UP; D = DOWN)	NA		NA
10				
11	**SUMMARY OF KEY RESULTS**	**2008**	**2009**	**2010**
12	NET INCOME AFTER TAXES	NA		
13	END-OF-THE-YEAR CASH ON HAND	NA		
14				
15	**CALCULATIONS**	**2008**	**2009**	**2010**
16	NUMBER OF UNITS SOLD IN A DAY	1000		
17	SELLING PRICE PER UNIT	7.00		
18	COST OF GOODS SOLD PER UNIT	3.00		
19	NUMBER OF UNITS SOLD IN A YEAR	NA		

FIGURE C-1   Tutorial skeleton 1

	A	B	C	D
21	**INCOME STATEMENT AND CASH FLOW STATEMENT**	**2008**	**2009**	**2010**
22	BEGINNING-OF-THE-YEAR CASH ON HAND	NA		
23				
24	SALES (REVENUE)	NA		
25	COST OF GOODS SOLD	NA		
26	INCOME BEFORE TAXES	NA		
27	INCOME TAX EXPENSE	NA		
28	NET INCOME AFTER TAXES	NA		
29				
30	END-OF-THE-YEAR CASH ON HAND (BEGINNING-OF-THE-YEAR CASH, PLUS NET INCOME AFTER TAXES)	10000		

FIGURE C-2   Tutorial skeleton 2

Each spreadsheet section is discussed next.

The CONSTANTS Section

This section of Figure C-1 records values that are used in spreadsheet calculations. In a sense, the constants are inputs, except that they do not change. In this tutorial, constants are TAX RATE and the NUMBER OF BUSINESS DAYS.

The INPUTS Section

The inputs shown in Figure C-1 are for the ECONOMIC OUTLOOK and PURCHASE-PRICE OUTLOOK (manufacturing input costs). Inputs could conceivably be entered for *each year* the model is covered (here, 2009 and 2010). That would let you enter an *O* for 2009's economy in one cell and a *P* for 2010's economy in another cell. Alternatively, one input for the two-year period could be entered in one cell. For simplicity, this tutorial uses the latter approach.

The SUMMARY OF KEY RESULTS Section

This section of the spreadsheet captures 2009 and 2010 NET INCOME AFTER TAXES (profit) for the year and END-OF-THE-YEAR CASH ON HAND, which you should assume are the two relevant outputs of this model. The summary merely repeats, in one easy-to-see place, results that appear in otherwise widely spaced places in the spreadsheet. (It also makes for easier charting at a later time.)

The CALCULATIONS Section

This area is used to compute the following data:

- The NUMBER OF UNITS SOLD IN A DAY, which is a function of the 2008 value and the economic outlook input
- The SELLING PRICE PER UNIT, which is similarly derived
- The COST OF GOODS SOLD PER UNIT, which is a function of the 2008 value and of purchase-price outlook
- The NUMBER OF UNITS SOLD IN A YEAR, which equals the number of units sold in a day times the number of business days

Those formulas could be embedded in the **INCOME STATEMENT AND CASH FLOW STATEMENT** section of the spreadsheet, which will be described shortly. Doing that, however, would result in expressions that are complex and difficult to understand. Putting the intermediate calculations into a separate **CALCULATIONS** section breaks up the work into modules. That is good form because it simplifies your programming.

The INCOME STATEMENT AND CASH FLOW STATEMENT Section

This section is the "body" of the spreadsheet. It shows the following:

- BEGINNING-OF-THE-YEAR CASH ON HAND, which equals cash at the end of the *prior* year.
- SALES (REVENUE), which equals the units sold in the year times the unit selling price.
- COST OF GOODS SOLD, which is units sold in the year times the price paid to acquire or make the unit sold.
- INCOME BEFORE TAXES, which equals sales, less total costs.
- INCOME TAX EXPENSE, which is zero when there are losses; otherwise, it is the income before taxes times the tax rate. (INCOME TAX EXPENSE is sometimes called INCOME TAXES.)
- NET INCOME AFTER TAXES, which equals income before taxes, less income tax expense.
- END-OF-THE-YEAR CASH ON HAND, which is beginning-of-the-year cash on hand plus net income. (In the real world, cash flow estimates must account for changes in receivables and payables. In this case, assume sales are collected immediately—that is, there are no receivables or bad debts. Also assume that suppliers are paid immediately—that is, there are no payables.)

Construction of the Spreadsheet Model

Next, you will work through the following three steps to build your spreadsheet model:

1. Make a "skeleton" of the spreadsheet and call it **TUTC.xlsx**.
2. Fill in the "easy" cell formulas.
3. Enter the "hard" spreadsheet formulas.

Make a Skeleton

Your first step is to set up a skeleton worksheet. The worksheet should have headings, text string labels, and constants—but no formulas.

To set up the skeleton, you must first grasp the problem *conceptually*. The best way to do that is to work *backward* from what the "body" of the spreadsheet will look like. Here the body is the **INCOME STATEMENT AND CASH FLOW STATEMENT** section. Set that up in your mind or on paper; then do the following:

- Decide what amounts should be in the **CALCULATIONS** section. In the income statement of this tutorial's model, SALES (revenue) will be NUMBER OF UNITS SOLD IN A DAY times SELLING PRICE PER UNIT. You will calculate the intermediate amounts (NUMBER OF UNITS SOLD IN A YEAR and SELLING PRICE PER UNIT) in the **CALCULATIONS** section.
- Set up the **SUMMARY OF KEY RESULTS** section by deciding what *outputs* are needed to solve the problem. The **INPUTS** section should be reserved for amounts that can change—the controlling variables—which are the ECONOMIC OUTLOOK and the PURCHASE-PRICE OUTLOOK.
- Use the **CONSTANTS** section for values you will need to use but are not in doubt; that is, you will not have to input them or calculate them. Here the TAX RATE is a good example of such a value.

⌨ AT THE KEYBOARD

Type in the Excel skeleton shown in Figure C-1 and Figure C-2.

> **NOTE**
>
> A designation of *NA* means that a cell will not be used in any formula in the worksheet. The 2008 values are needed only for certain calculations; so for the most part, the 2008 column's cells show *NA*. (Recall that the forecast is for 2009 and 2010.) Also be aware that you can "break" a text string in a cell by pressing the Alt and Enter keys at the same time at the break point. That makes the cell "taller." Formatting of cells to show centered data and creation of borders is discussed at the end of this tutorial.

Fill in the "Easy" Formulas

The next step in building a spreadsheet model is to fill in the "easy" formulas. The cells affected (and what you should enter) are discussed next.

To prepare, you should format the cells in the **SUMMARY OF KEY RESULTS** section for no decimals. (Formatting for numerical precision is discussed at the end of this tutorial.) The **SUMMARY OF KEY RESULTS** section is shown in Figure C-3. As previously mentioned, the **SUMMARY OF KEY RESULTS** section simply echoes results shown in other places. Consider Figure C-1 and Figure C-2. Note that C28 in Figure C-2 holds the NET INCOME AFTER TAXES. The idea here is for you to echo that amount in C12 of Figure C-1. So the formula in C12 is =C28. Translation: "Copy what is in C28 into C12." It's that simple.

> **NOTE**
>
> With the insertion point in C12, the contents of that cell—in this case, the formula =C28—show in the editing window, which is above the lettered column indicators, as shown in Figure C-3.

At this point, C28 is empty (and thus has a zero value), but that does not prevent you from copying. So copy cell C12's formula to the right, to cell D12. Copying puts =D28 into D12, which is what you want. (Year 2010's NET INCOME AFTER TAXES is in D28.)

	A	B	C	D
	C12	▼	*fx*	=C28
11	SUMMARY OF KEY RESULTS	2008	2009	2010
12	NET INCOME AFTER TAXES	NA	0	
13	END-OF-THE-YEAR CASH ON HAND	NA		

FIGURE C-3   Echo 2009 NET INCOME AFTER TAXES

To perform the Copy operation, use the following steps:

1. Select (click in) the cell (or range of cells) that you want to copy. That activates the cell (or range) for copying.
2. Press the Control key and *C* at the same time (Ctrl+C).
3. Select the destination cell. (If a range of cells is the destination, select the upper-left cell of the destination range.)
4. Press the Control key and *V* at the same time (Ctrl+V).
5. Press the Escape key to deactivate the copied cell (or range).

Another way to copy is as follows:

1. Select the Home tab.
2. Select (click in) the cell (or range of cells) that you want to copy.
3. In the Clipboard group, select Copy.
4. Select the destination cell. (If a range of cells is the destination, select the upper-left cell of the destination range.)
5. In the Clipboard group, select Paste.
6. Press the Escape key to deactivate the copied cell (or range).

As you can see in Figure C-4, END-OF-THE-YEAR CASH ON HAND for 2009 cash is in cell C13. Echo the cash results in cell C30 to cell C13. (Put =C30 in cell C13, as shown in Figure C-4.) Copy the formula from C13 to D13.

	A	B	C	D
	C13	▼	*fx*	=C30
11	SUMMARY OF KEY RESULTS	2008	2009	2010
12	NET INCOME AFTER TAXES	NA	0	0
13	END-OF-THE-YEAR CASH ON HAND	NA	0	

FIGURE C-4   Echo 2009 END-OF-THE-YEAR CASH ON HAND

At this point, the **CALCULATIONS** section formulas will not be entered because they are not all "easy" formulas. Move on to the easier formulas in the **INCOME STATEMENT AND CASH FLOW STATEMENT** section, as if the calculations were already done. Again, the fact that the **CALCULATIONS** section cells are empty does not stop you from entering formulas in this section. You should format the cells in the **INCOME STATEMENT AND CASH FLOW STATEMENT** section for zero decimals.

As you can see in Figure C-5, BEGINNING-OF-THE-YEAR CASH ON HAND is the cash on hand at the end of the *prior* year. In C22 for 2009, type =B30. The "skeleton" you just entered is shown in Figure C-5. Cell B30 has the END-OF-THE-YEAR CASH ON HAND for 2008.

Figure C-6 shows the next step, which is to copy the formula in cell C22 to the right. SALES (REVENUE) is NUMBER OF UNITS SOLD IN A YEAR times SELLING PRICE PER UNIT. In cell C24, enter =C17*C19, as shown in Figure C-6.

The formula C17*C19 multiplies units sold for the year by the unit selling price. (Cells C17 and C19 are empty now, which is why SALES shows a zero after the formula is entered.) Copy the formula to the right to D24.

COST OF GOODS SOLD is handled similarly. In C25, type =C18*C19. That equals NUMBER OF UNITS SOLD IN A YEAR times COST OF GOODS SOLD PER UNIT. Copy the formula to the right.

In cell C26, the formula for INCOME BEFORE TAXES is =C24–C25. Enter the formula. Copy it to the right.

C22		f_x =B30		
	A	B	C	D
21	**INCOME STATEMENT AND CASH FLOW STATEMENT**	**2008**	**2009**	**2010**
22	BEGINNING-OF-THE-YEAR CASH ON HAND	NA	10000	
23				
24	SALES (REVENUE)	NA		
25	COST OF GOODS SOLD	NA		
26	INCOME BEFORE TAXES	NA		
27	INCOME TAX EXPENSE	NA		
28	NET INCOME AFTER TAXES	NA		
29				
30	END-OF-THE-YEAR CASH ON HAND (BEGINNING-OF-THE-YEAR CASH, PLUS NET INCOME AFTER TAXES)	10000		

FIGURE C-5   Echo of END-OF-THE-YEAR CASH ON HAND for 2008 to BEGINNING-OF-THE-YEAR CASH ON HAND for 2009

C24		f_x =C17*C19		
	A	B	C	D
15	**CALCULATIONS**	**2008**	**2009**	**2010**
16	NUMBER OF UNITS SOLD IN A DAY	1000		
17	SELLING PRICE PER UNIT	7.00		
18	COST OF GOODS SOLD PER UNIT	3.00		
19	NUMBER OF UNITS SOLD IN A YEAR	NA		
20				
21	**INCOME STATEMENT AND CASH FLOW STATEMENT**	**2008**	**2009**	**2010**
22	BEGINNING-OF-THE-YEAR CASH ON HAND	NA	10000	0
23				
24	SALES (REVENUE)	NA	0	
25	COST OF GOODS SOLD	NA		
26	INCOME BEFORE TAXES	NA		
27	INCOME TAX EXPENSE	NA		
28	NET INCOME AFTER TAXES	NA		
29				
30	END-OF-THE-YEAR CASH ON HAND (BEGINNING-OF-THE-YEAR CASH, PLUS NET INCOME AFTER TAXES)	10000		

FIGURE C-6   Enter the formula to compute 2009 SALES

In the United States, income taxes are paid only on positive income before taxes. In cell C27, the INCOME TAX EXPENSE is zero when the INCOME BEFORE TAXES is zero or less; otherwise, INCOME TAX EXPENSE equals the TAX RATE times the INCOME BEFORE TAXES. The TAX RATE is a constant (in C4). An IF statement is needed to express this logic:

IF(INCOME BEFORE TAXES is <= 0, put zero tax in C27,
else in C27 put a number equal to multiplying the
TAX RATE times the INCOME BEFORE TAXES)

C26 stands for the concept INCOME BEFORE TAXES, and C4 stands for the concept TAX RATE. So in Excel, substitute those cell addresses:

=IF(C26 <= 0, 0, C4 * C26)

Copy the income tax expense formula to the right.

In cell C28, NET INCOME AFTER TAXES is INCOME BEFORE TAXES, less INCOME TAX EXPENSE: =C26–C27. Enter and copy the formula to the right.

The END-OF-THE-YEAR CASH ON HAND is BEGINNING-OF-THE-YEAR CASH ON HAND plus NET INCOME AFTER TAXES. In cell C30, enter =C22+C28. The **INCOME STATEMENT AND CASH FLOW STATEMENT** section at that point is shown in Figure C-7. Copy the formula to the right.

	C30		f_x	=C22+C28	
	A	B	C	D	

	A	B	C	D
21	INCOME STATEMENT AND CASH FLOW STATEMENT	2008	2009	2010
22	BEGINNING-OF-THE-YEAR CASH ON HAND	NA	10000	10000
23				
24	SALES (REVENUE)	NA	0	0
25	COST OF GOODS SOLD	NA	0	0
26	INCOME BEFORE TAXES	NA	0	0
27	INCOME TAX EXPENSE	NA	0	0
28	NET INCOME AFTER TAXES	NA	0	0
29				
30	END-OF-THE-YEAR CASH ON HAND (BEGINNING-OF-THE-YEAR CASH, PLUS NET INCOME AFTER TAXES)	10000	10000	

FIGURE C-7   Status of INCOME STATEMENT AND CASH FLOW STATEMENT

Put in the "Hard" Formulas

The next step is to finish the spreadsheet by filling in the "hard" formulas.

■ AT THE KEYBOARD

In C8, enter an *O* for OPTIMISTIC; and in C9, enter *U* for UP. There is nothing magical about those particular values—they just give the worksheet formulas some input to process. Recall that the inputs will cover both 2009 and 2010. Make sure *NA* is in D8 and D9 just to remind yourself that those cells will not be used for input or by other worksheet formulas. Your **INPUTS** section should look like the one shown in Figure C-8.

	A	B	C	D
7	INPUTS	2008	2009	2010
8	ECONOMIC OUTLOOK (O = OPTIMISTIC; P = PESSIMISTIC)	NA	O	NA
9	PURCHASE-PRICE OUTLOOK (U = UP; D = DOWN)	NA	U	NA

FIGURE C-8   Entering two input values

Recall that cell addresses in the **CALCULATIONS** section are already referred to in formulas in the **INCOME STATEMENT AND CASH FLOW STATEMENT** section. The next step is to enter formulas for those calculations. Before doing that, format NUMBER OF UNITS SOLD IN A DAY and NUMBER OF UNITS SOLD IN A YEAR for zero decimals and format SELLING PRICE PER UNIT and COST OF GOODS SOLD PER UNIT for two decimals.

The easiest formula in the **CALCULATIONS** section is the NUMBER OF UNITS SOLD IN A YEAR, which is the calculated NUMBER OF UNITS SOLD IN A DAY (in C16) times the NUMBER OF BUSINESS DAYS (in C5). In C19, enter =C5*C16, as shown in Figure C-9.

	A	B	C	D
	C19		f_x =C5*C16	
1	**TUTORIAL EXERCISE**			
2				
3	CONSTANTS	2008	2009	2010
4	TAX RATE	NA	0.33	0.35
5	NUMBER OF BUSINESS DAYS	NA	300	300
6				
7	INPUTS	2008	2009	2010
8	ECONOMIC OUTLOOK (O = OPTIMISTIC; P = PESSIMISTIC)	NA	O	NA
9	PURCHASE-PRICE OUTLOOK (U = UP; D = DOWN)	NA	U	NA
10				
11	SUMMARY OF KEY RESULTS	2008	2009	2010
12	NET INCOME AFTER TAXES	NA	0	0
13	END-OF-THE-YEAR CASH ON HAND	NA	10000	10000
14				
15	CALCULATIONS	2008	2009	2010
16	NUMBER OF UNITS SOLD IN A DAY	1000		
17	SELLING PRICE PER UNIT	7.00		
18	COST OF GOODS SOLD PER UNIT	3.00		
19	NUMBER OF UNITS SOLD IN A YEAR	NA	0	

FIGURE C-9   Entering the formula to compute 2009 NUMBER OF UNITS SOLD IN A YEAR

Copy the formula to cell D19 for year 2010.

Assume that if the ECONOMIC OUTLOOK is OPTIMISTIC, the 2009 NUMBER OF UNITS SOLD IN A DAY will be 6 percent more than that in 2008; in 2010, they will be 6 percent more than that in 2009. Also assume that if the ECONOMIC OUTLOOK is PESSIMISTIC, the NUMBER OF UNITS SOLD IN A DAY in 2009 will be 1 percent less than those sold in 2008; in 2010, they will be 1 percent less than those sold in 2009. An IF statement is needed in C16 to express that idea:

> IF(economy variable = OPTIMISTIC,
>> then NUMBER OF UNITS SOLD IN A DAY will go UP 6%,
>>> else NUMBER OF UNITS SOLD IN A DAY will go DOWN 1%)

Substituting cell addresses:

> =IF(C8 = "O", 1.06 * B16, 0.99 * B16)

NOTE

In Excel, quotation marks denote labels. The input is a one-letter label. So the quotation marks around the O are needed. Also note that multiplying by 1.06 results in a 6 percent increase, whereas multiplying by .99 results in a 1 percent decrease.

Enter the entire IF formula into cell C16, as shown in Figure C-10. Absolute addressing is needed (C8), because the address is in a formula that gets copied *and* you do not want the cell reference to change (to D8, which has the value *NA*) when you copy the formula to the right. Absolute addressing maintains the C8 reference when the formula is copied. Copy the formula in C16 to D16 for 2010.

	A	B	C	D
	C16		f_x =IF(C8="O",1.06*B16,0.99*B16)	
15	CALCULATIONS	2008	2009	2010
16	NUMBER OF UNITS SOLD IN A DAY	1000	1060	
17	SELLING PRICE PER UNIT	7.00		
18	COST OF GOODS SOLD PER UNIT	3.00		

FIGURE C-10   Entering the formula to compute 2009 NUMBER OF UNITS SOLD IN A DAY

The SELLING PRICE PER UNIT is also a function of the ECONOMIC OUTLOOK. Assume the two-part rule is as follows:

- If the ECONOMIC OUTLOOK is OPTIMISTIC, the SELLING PRICE PER UNIT in 2009 will be 1.07 times that of 2008; in 2010, it will be 1.07 times that of 2009.
- On the other hand, if the ECONOMIC OUTLOOK is PESSIMISTIC, the SELLING PRICE PER UNIT in 2009 and 2010 will equal the per-unit price in 2008; that is, the price will not change.

Test your understanding of the selling price calculation by figuring out the formula for cell C17. Enter the formula and copy it to the right. You will need to use absolute addressing. (Can you see why?)

The COST OF GOODS SOLD PER UNIT is a function of the PURCHASE-PRICE OUTLOOK. Assume the two-part rule is as follows:

- If the PURCHASE-PRICE OUTLOOK is UP (*U*), COST OF GOODS SOLD PER UNIT in 2009 will be 1.25 times that of year 2008; in 2010, it will be 1.25 times that of 2009.
- On the other hand, if the PURCHASE-PRICE OUTLOOK is DOWN (*D*), the multiplier each year will be 1.01.

Again, to test your understanding, figure out the formula for cell C18. Enter and copy the formula to the right. You will need to use absolute addressing.

Your selling price and cost of goods sold formulas, given OPTIMISTIC and UP input values, should yield the calculated values shown in Figure C-11.

	A	B	C	D
15	CALCULATIONS	2008	2009	2010
16	NUMBER OF UNITS SOLD IN A DAY	1000	1060	1124
17	SELLING PRICE PER UNIT	7.00	7.49	8.01
18	COST OF GOODS SOLD PER UNIT	3.00	3.75	4.69
19	NUMBER OF UNITS SOLD IN A YEAR	NA	318000	337080

FIGURE C-11   Calculated values given OPTIMISTIC and UP input values

Assume you change the input values to PESSIMISTIC and DOWN. Your formulas should yield the calculated values shown in Figure C-12.

	A	B	C	D
15	CALCULATIONS	2008	2009	2010
16	NUMBER OF UNITS SOLD IN A DAY	1000	990	980
17	SELLING PRICE PER UNIT	7.00	7.00	7.00
18	COST OF GOODS SOLD PER UNIT	3.00	3.03	3.06
19	NUMBER OF UNITS SOLD IN A YEAR	NA	297000	294030

FIGURE C-12   Calculated values given PESSIMISTIC and DOWN input values

That completes the body of your spreadsheet. The values in the **CALCULATIONS** section ripple through the **INCOME STATEMENT AND CASH FLOW STATEMENT** section because the income statement formulas reference the calculations. Assuming inputs of OPTIMISTIC and UP, the income and cash flow numbers should look like those in Figure C-13.

	A	B	C	D
21	INCOME STATEMENT AND CASH FLOW STATEMENT	2008	2009	2010
22	BEGINNING-OF-THE-YEAR CASH ON HAND	NA	10000	806844
23				
24	SALES (REVENUE)	NA	2381820	2701460
25	COST OF GOODS SOLD	NA	1192500	1580063
26	INCOME BEFORE TAXES	NA	1189320	1121398
27	INCOME TAX EXPENSE	NA	392476	392489
28	NET INCOME AFTER TAXES	NA	796844	728909
29				
30	END-OF-THE-YEAR CASH ON HAND (BEGINNING-OF-THE-YEAR CASH, PLUS NET INCOME AFTER TAXES)	10000	806844	1535753

FIGURE C-13   Completed INCOME STATEMENT AND CASH FLOW STATEMENT section

SCENARIO MANAGER

You are now ready to use the Scenario Manager to capture inputs and results as you play "what if" with the spreadsheet.

Note that there are four possible combinations of input values: O-U (Optimistic-Up), O-D (Optimistic-Down), P-U (Pessimistic-Up), and P-D (Pessimistic-Down). Financial results for each combination will be different. Each combination of input values can be referred to as a "scenario." Excel's Scenario Manager records the results of each combination of input values as a separate scenario and then shows a summary of all scenarios in a separate worksheet. Those summary worksheet values can be used as a raw table of numbers and then printed or copied into a Word document. The table of data can then be the basis for an Excel chart, which also can be printed or inserted into a memo.

You have four possible scenarios for the economy and the purchase price of goods sold: Optimistic-Up, Optimistic-Down, Pessimistic-Up, and Pessimistic-Down. The four input-value sets produce different financial results. When you use the Scenario Manager, define the four scenarios; then have Excel (1) sequentially run the input values "behind the scenes" and (2) put the results for each input scenario in a summary sheet.

When you define a scenario, you give the scenario a name and identify the input cells and input values. You do that for each scenario. Then you identify the output cells so Excel can capture the outputs in a summary sheet.

AT THE KEYBOARD

To start, select the Data tab. The Data Tools group has a What-If Analysis icon. Select the drop-down arrow and click the Scenario Manager menu choice. Initially, no scenarios are defined, and Excel tells you that, as you can see in Figure C-14.

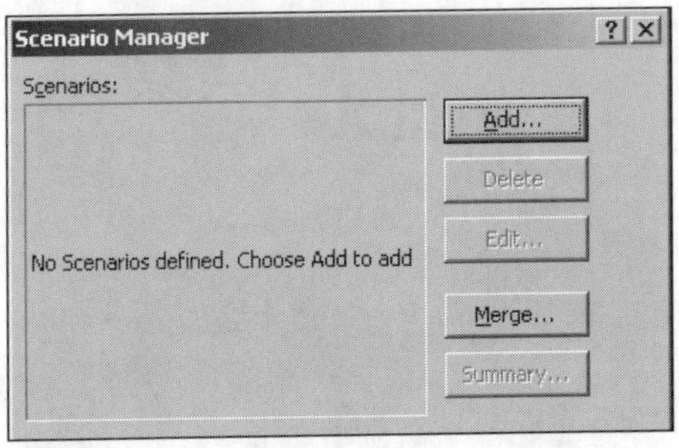

FIGURE C-14   Initial Scenario Manager window

With this window, you are able to add, delete, or change (edit) scenarios. Toward the end of the process, you also can create the summary sheet.

NOTE

When working with this window and its successors, do not press the Enter key to navigate. Use mouse clicks to move from one step to the next.

To continue with defining a scenario, click the Add button. In the resulting Add Scenario window, give the first scenario a name: OPT-UP. Then type the input cells in the Changing cells field—here, C8:C9. (*Note:* C8 and C9 are contiguous input cells. Noncontiguous input cell ranges are separated by a comma.) Excel may add dollar signs to the cell address, but do not be concerned about that. The window should look like the one shown in Figure C-15.

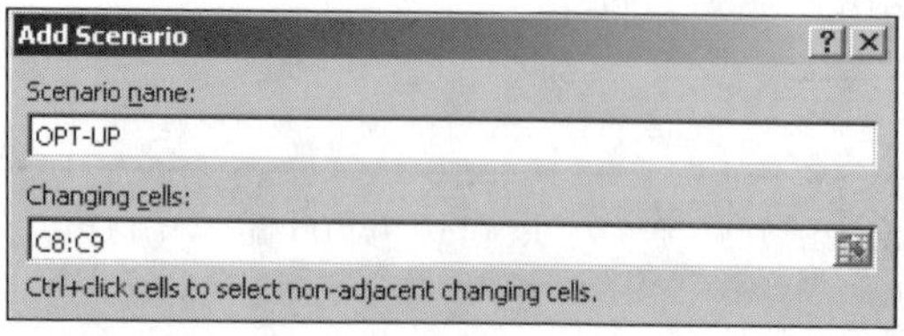

FIGURE C-15 Entering OPT-UP as a scenario

Now click **OK**, which moves you to the Scenario Values window. Here you indicate what the INPUT *values* will be for the scenario. The values in the cells *currently* in the spreadsheet will be displayed. They might or might not be correct for the scenario you are defining. For the OPT-UP scenario, you need to enter an *O* and a *U*, if not the current values. Enter those values if needed, as shown in Figure C-16.

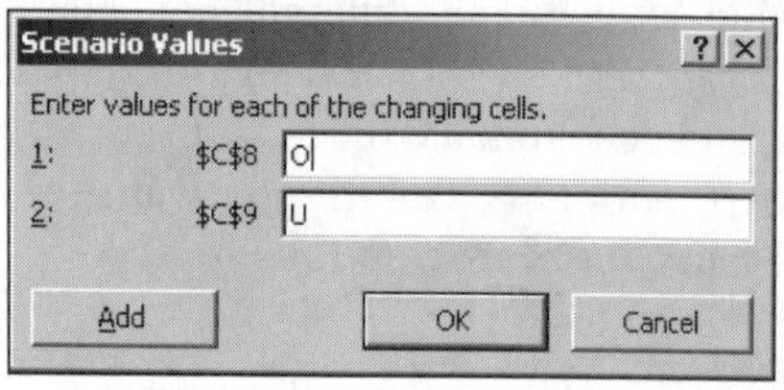

FIGURE C-16 Entering OPT-UP scenario input values

Click **OK**, which takes you back to the Scenario Manager window. You can now enter the other three scenarios, following the same steps. Do that now. Enter the OPT-DOWN, PESS-UP, and PESS-DOWN scenarios and related input values. After doing that, you should see that the names and changing cells for the four scenarios have been entered, as in Figure C-17.

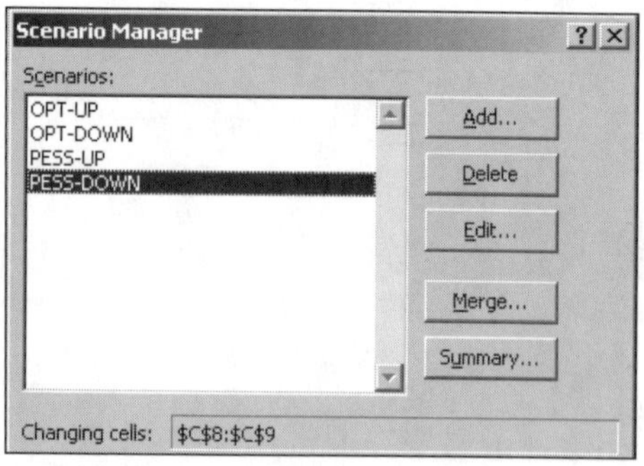

FIGURE C-17 Scenario Manager window with all scenarios entered

You are now able to create a summary sheet that shows the results of running the four scenarios. Click the Summary button. You'll get the Scenario Summary window. You must provide Excel with the output cell addresses—they will be the same for all four scenarios. (The output *values* in those output cells change as input values are changed, but the addresses of the output cells do not change.)

Assume you are interested primarily in the results that have accrued at the end of the two-year period. Those are your two 2010 **SUMMARY OF KEY RESULTS** section cells for NET INCOME AFTER TAXES and END-OF-THE-YEAR CASH ON HAND (D12 and D13). Type those addresses in the window's input area, as shown in Figure C-18. (*Note:* If result cells are noncontiguous, the address ranges can be entered, separated by a comma.)

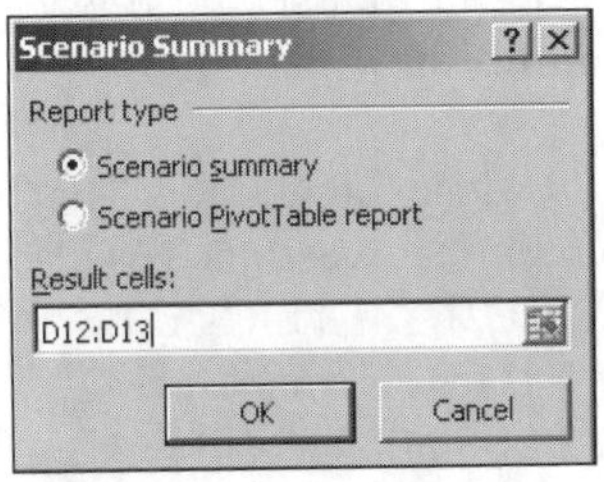

FIGURE C-18 Entering Result cells addresses in Scenario Summary window

Then click **OK**. Excel runs each set of inputs and collects results as it goes. (You do not see that happening on the screen.) Excel makes a *new* sheet titled the "Scenario Summary" (denoted by the sheet's lower tab) and takes you there, as shown in Figure C-19.

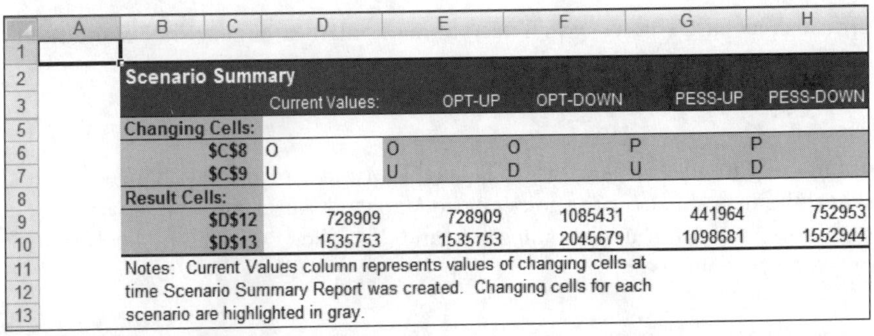

	Current Values:	OPT-UP	OPT-DOWN	PESS-UP	PESS-DOWN
Scenario Summary					
Changing Cells:					
C8	O	O	O	P	P
C9	U	U	D	U	D
Result Cells:					
D12	728909	728909	1085431	441964	752953
D13	1535753	1535753	2045679	1098681	1552944

Notes: Current Values column represents values of changing cells at time Scenario Summary Report was created. Changing cells for each scenario are highlighted in gray.

FIGURE C-19 Scenario Summary sheet created by Scenario Manager

One somewhat annoying visual element is that the Current Values in the spreadsheet are given an output column. That duplicates one of the four defined scenarios. You can delete that extra column by following these steps: (1) Select the Home tab; (2) within the Cells group, select the Delete icon's drop-down arrow; and (3) select Delete Sheet Columns.

NOTE

A sheet's row can be deleted the same way as a column can be deleted. You select Delete Sheet Rows at the end.

Another annoyance is that column A goes unused. You can click and delete it as you've been doing to move everything to the left. That should make columns of data easier to see on the screen without scrolling. Other ways to make the worksheet easier to read include: (1) entering words in column A to describe the input and output cells; (2) centering cell values by using the Center Text icon in the Home tab's Alignment group; and (3) showing data in Currency format by using the Number Format drop-down menu within the Home tab's Number group.

When you finish, your summary sheet should resemble the one shown in Figure C-20.

	A	B	C	D	E	F
1	Scenario Summary					
2			OPT-UP	OPT-DOWN	PESS-UP	PESS-DOWN
4	Changing Cells:					
5	ECONOMIC OUTLOOK	C8	O	O	P	P
6	PURCHASE PRICE OUTLOOK	C9	U	D	U	D
7	Result Cells (2010):					
8	NET INCOME AFTER TAXES	D12	$728,909	$1,085,431	$441,964	$752,953
9	END-OF-THE-YEAR CASH ON HAND	D13	$1,535,753	$2,045,679	$1,098,681	$1,552,944

FIGURE C-20   Scenario Summary sheet after formatting

Note that column C shows the OPTIMISTIC-UP case. NET INCOME AFTER TAXES in that scenario is $728,909, and END-OF-THE-YEAR CASH ON HAND is $1,535,753. Columns D, E, and F show the other scenario results.

Here is an important postscript to this exercise: DSS spreadsheets are used to guide decision making. That means that the spreadsheet's results must be interpreted in some way. Here are two practice questions based on the results in Figure C-20: With that data, what combination of year 2010 NET INCOME AFTER TAXES and END-OF-THE-YEAR CASH ON HAND would be best? Clearly, OPTIMISTIC-DOWN (O-D) is the best result, right? It yields the highest income and highest cash. What is the worst combination? PESSIMISTIC-UP (P-U), right? It yields the lowest income and lowest cash.

Results are not always that easy to interpret, but the analytical method is the same. You have a complex situation that you cannot understand very well without software assistance. You build a model of the situation in the spreadsheet, enter the inputs, collect the results, and then interpret the results to help with decision making.

Summary Sheets

When you do Scenario Manager spreadsheet case studies, you'll need to manipulate summary sheets and their data. Now you will look at some of those operations.

Rerunning the Scenario Manager

The Scenario Summary sheet does not update itself when the spreadsheet formulas or inputs change. To see an updated Scenario Summary sheet, the user must rerun the Scenario Manager. To rerun the Scenario Manager, click the Summary button in the Scenario Manager dialog box, and then click **OK**. That makes another summary sheet. It does not overwrite a prior one.

Deleting Unwanted Scenario Manager Summary Sheets

Suppose you want to delete a summary sheet. With the summary sheet on the screen, follow this procedure: (1) Select the Home tab; (2) within the Cells group, select the Delete icon's drop-down arrow; and (3) select Delete Sheet. Note that you will be asked if you really mean to delete this sheet. Click Delete if you do.

Charting Summary Sheet Data

The summary sheet results can be conveniently charted using the Chart Wizard. Charting Excel data is discussed in Tutorial E.

Copying Summary Sheet Data to the Clipboard

You may want to put the summary sheet data into the Clipboard to use in a Word document. To do that, follow these steps:

1. Select the data range.
2. Copy the data range by following the copying operation described earlier in this tutorial. That puts the data range into the clipboard.
3. Open your Word document.
4. Click the cursor where you want the upper-left part of the graphic to be positioned.
5. Paste the data into the document by selecting Paste in the Home tab's Clipboard group.

PRACTICE USING SCENARIO MANAGER

Suppose you have an uncle who works for a large company. He has a good job and makes a decent salary (currently, $80,000 a year). At age 65, he can retire from his company, which will be in 2015. He will start drawing his pension then.

However, the company has an early-out plan. Under the plan, the company asks employees to quit (called "preretirement"). The company then pays those employees a bonus in the year they quit and each year thereafter, up to the official retirement date, which is through the year 2014 for your uncle. Then employees start to receive their actual pension—in your uncle's case, in 2015. This early-out program would let your uncle leave the company before 2015. Until then, he could find a part-time hourly job to make ends meet and then leave the workforce entirely in 2015.

The opportunity to leave early is open through 2014. That means your uncle could stay with the company in 2009, then leave the company any year in the period 2010 to 2014 and get the early-out bonuses in the years he is retired. If he retires in 2010, he would lose the 2009 bonus. If he retires in 2011, he would lose the 2009 and 2010 bonuses. That would continue all the way through 2014.

Another factor in your uncle's thinking is whether to continue his country club membership. He likes the club, but it is a cash drain. The early-out decision can be assessed each year, but the decision about the country club membership must be made now; if your uncle does not withdraw in 2009, he says he will remain a member (and incur costs) through 2014.

Your uncle has called you in to make a Scenario Manager spreadsheet model of his situation. Your spreadsheet would let him play "what if" with the preretirement and country club possibilities and see his various projected personal finance results for 2009–2014. With each scenario, your uncle wants to know what "cash on hand" will be available for each year in the period.

Complete the spreadsheet for your uncle. Your **SUMMARY OF KEY RESULTS, CALCULATIONS**, and **INCOME STATEMENT AND CASH FLOW STATEMENT** section cells must show values *by cell formula*. That is, in those areas, do not hard-code amounts. Also do not use the address of a cell if its contents are *NA* in any of your formulas. Set up your spreadsheet skeleton as shown in the figures that follow. Name your spreadsheet **UNCLE.xlsx**.

CONSTANTS Section

Your spreadsheet should have the constants shown in Figure C-21. An explanation of line items follows the figure.

	A	B	C	D	E	F	G	H
1	**YOUR UNCLE'S EARLY RETIREMENT DECISION**							
2	CONSTANTS	2008	2009	2010	2011	2012	2013	2014
3	SALARY INCREASE FACTOR	NA	0.03	0.03	0.02	0.02	0.01	0.01
4	PART TIME WAGES EXPECTED	NA	10000	10200	10500	10800	11400	12000
5	BUY OUT AMOUNT	NA	30000	25000	20000	15000	5000	0
6	COST OF LIVING (NOT RETIRED)	NA	41000	42000	43000	44000	45000	46000
7	COUNTRY CLUB DUES	NA	12000	13000	14000	15000	16000	17000

FIGURE C-21   CONSTANTS section values

- SALARY INCREASE FACTOR: Your uncle's salary at the end of 2008 will be $80,000. As you can see, raises are expected each year; for example, a 3 percent raise is expected in 2009. If your uncle does not retire, he will get his salary and small raise for the year.
- PART TIME WAGES EXPECTED: Your uncle has estimated his part-time wages as if he were retired and working part-time in the 2009–2014 period.
- BUY OUT AMOUNT: The amounts for the company's preretirement buyout plan are shown. If your uncle retires in 2009, he gets $30,000, $25,000, $20,000, $15,000, $5,000, and zero in the years 2009 to 2014, respectively. If he leaves in 2010, he will give up the $30,000 payment for 2009, but will get $25,000, $20,000, $15,000, $5,000, and zero in the years 2010 to 2014, respectively.
- COST OF LIVING (NOT RETIRED): Your uncle has estimated how much cash he needs to meet his living expenses, assuming he continues to work for the company. His cost of living would be $41,000 in 2009, increasing each year thereafter.
- COUNTRY CLUB DUES: Country club dues are $12,000 for 2009. They increase each year thereafter.

INPUTS Section

Your spreadsheet should have the inputs shown in Figure C-22. An explanation of line items follows the figure.

	A	B	C	D	E	F	G	H
9	INPUTS	2008	2009	2010	2011	2012	2013	2014
10	RETIRED [R] or WORKING [W]	NA						
11	STAY IN CLUB? [Y] OR [N]	NA		NA	NA	NA	NA	NA

FIGURE C-22   INPUTS section

- RETIRED OR WORKING: Enter an **R** if your uncle retires in a year or a **W** if he is still working. If he is working through 2014, you should enter the pattern **WWWWWW**. If his retirement is in 2009, you should enter the pattern **RRRRRR**. If he works for three years and then retires in 2012, you should enter the pattern **WWWRRR**.
- STAY IN CLUB?: If your uncle stays in the club from 2009–2014, you should enter a **Y**. If your uncle leaves the club in 2009, you should enter an **N**. The decision applies to all years.

SUMMARY OF KEY RESULTS Section

Your spreadsheet should show the results in Figure C-23.

	A	B	C	D	E	F	G	H
13	SUMMARY OF KEY RESULTS	2008	2009	2010	2011	2012	2013	2014
14	END-OF-THE-YEAR CASH ON HAND	NA						

FIGURE C-23   SUMMARY OF KEY RESULTS section

Each year's END-OF-THE-YEAR CASH ON HAND value is echoed from cells in the spreadsheet body.

CALCULATIONS Section

Your spreadsheet should calculate, by formula, the values shown in Figure C-24. Calculated amounts are used later in the spreadsheet. An explanation of line items follows the figure.

	A	B	C	D	E	F	G	H
16	CALCULATIONS	2008	2009	2010	2011	2012	2013	2014
17	TAX RATE	NA						
18	COST OF LIVING	NA						
19	YEARLY SALARY OR WAGES	80000						
20	COUNTRY CLUB DUES PAID	NA						

FIGURE C-24   CALCULATIONS section

- TAX RATE: Your uncle's tax rate depends on whether he is retired. Retired people have lower overall tax rates. If he is retired in a year, your uncle's rate is expected to be 15 percent of income before taxes. In a year in which he works full-time, the rate will be 30 percent.
- COST OF LIVING: In any year that your uncle continues to work for the company, his cost of living is the amount shown in COST OF LIVING (NOT RETIRED) in the **CONSTANTS** section in Figure C-21. But if he chooses to retire, his cost of living is $15,000 less than the amount shown in the figure.
- YEARLY SALARY OR WAGES: If your uncle keeps working, his salary increases each year. The year-to-year percentage increases are shown in the **CONSTANTS** section. Thus, salary earned in 2009 would be more than that earned in 2008, salary earned in 2010 would be more than that earned in 2009, and so on. If your uncle retires in a certain year, he will make the part-time wages shown in the **CONSTANTS** section.
- COUNTRY CLUB DUES PAID: If your uncle leaves the club, the dues are zero each year; otherwise, the dues are as shown in the **CONSTANTS** section.

The INCOME STATEMENT AND CASH FLOW STATEMENT Section

This section begins with the cash on hand at the beginning of the year. That is followed by the income statement, concluding with the calculation of cash on hand at the end of the year. The format is shown in Figure C-25. An explanation of line items follows the figure.

	A	B	C	D	E	F	G	H
22	INCOME STATEMENT AND CASH FLOW STATEMENT	2008	2009	2010	2011	2012	2013	2014
23	BEGINNING-OF-THE-YEAR CASH ON HAND	NA						
24								
25	SALARY OR WAGES	NA						
26	BUY OUT INCOME	NA						
27	TOTAL CASH INFLOW	NA						
28	COUNTRY CLUB DUES PAID	NA						
29	COST OF LIVING	NA						
30	TOTAL COSTS	NA						
31	INCOME BEFORE TAXES	NA						
32	INCOME TAX EXPENSE	NA						
33	NET INCOME AFTER TAXES	NA						
34								
35	END-OF-THE-YEAR CASH ON HAND (BEGINNING-OF-THE-YEAR CASH, PLUS NET INCOME AFTER TAXES)	30000						

FIGURE C-25   INCOME STATEMENT AND CASH FLOW STATEMENT section

- BEGINNING-OF-THE-YEAR CASH ON HAND: This is the END-OF-THE-YEAR CASH ON HAND at the end of the prior year.
- SALARY OR WAGES: This is a yearly calculation, which can be echoed here.
- BUY OUT INCOME: This is the year's buyout amount if your uncle is retired that year.
- TOTAL CASH INFLOW: This is the sum of salary or part-time wages and buyout amounts.
- COUNTRY CLUB DUES PAID: This is a calculated amount, which can be echoed here.
- COST OF LIVING: This is a calculated amount, which can be echoed here.
- TOTAL COSTS: These outflows are the sum of the COST OF LIVING and COUNTRY CLUB DUES PAID.
- INCOME BEFORE TAXES: This amount is the total CASH inflow, less total costs (outflows).
- INCOME TAX EXPENSE: This amount is zero when INCOME BEFORE TAXES is zero or less; otherwise, the calculated tax rate is applied to the INCOME BEFORE TAXES.
- NET INCOME AFTER TAXES: This is income before taxes, less tax expense.
- END-OF-THE-YEAR CASH ON HAND: This is the BEGINNING-OF-THE-YEAR CASH ON HAND plus the year's NET INCOME AFTER TAXES.

Scenario Manager Analysis

Set up the Scenario Manager and create a Scenario Summary sheet. Your uncle wants to look at the following four possibilities:

- Retire in 2009, staying in the club ("Loaf-In")
- Retire in 2009, leaving the club ("Loaf-Out")
- Work three more years and retire in 2012, staying in the club ("Delay-In")
- Work three more years and retire in 2012, leaving the club ("Delay-Out")

You can enter noncontiguous cell ranges as follows: C20..F20, C21, C22 (cell addresses are examples). The output cell should be only the 2013 END-OF-THE-YEAR CASH ON HAND cell.

Your uncle will choose the option that yields the highest 2014 END-OF-THE-YEAR CASH ON HAND. You must look at your Scenario Summary sheet to see which strategy yields the highest amount.

To check your work, you should attain the values shown in Figure C-26. (You can use the labels that Excel gives you in the far left column or change the labels, as was done in Figure C-26.)

	A	B	C	D	E	F
1	Scenario Summary					
2			LOAF-IN	LOAF-OUT	DELAY-IN	DELAY-OUT
4	Changing Cells:					
5	RETIRE OR WORK, 2009	C10	R	R	W	W
6	RETIRE OR WORK, 2010	D10	R	R	W	W
7	RETIRE OR WORK, 2011	E10	R	R	W	W
8	RETIRE OR WORK, 2012	F10	R	R	R	R
9	RETIRE OR WORK, 2013	G10	R	R	R	R
10	RETIRE OR WORK, 2014	H10	R	R	R	R
11	IN CLUB, 2009-2014?	C11	Y	N	Y	N
12	Result Cells (2014):					
13	END-OF-THE-YEAR CASH ON HAND	H14	-$68,400	$15,195	$8,389	$83,689

FIGURE C-26   Scenario Summary

REVIEW OF EXCEL BASICS

In this section, you'll begin by reviewing how to perform some basic operations. Then you'll work through more cash flow calculations. Reading and working through this section will help you complete the spreadsheet cases in this book.

Basic Operations

To begin, you'll review the following topics: formatting cells, showing Excel cell formulas, understanding circular references, using the AND and the OR functions in IF statements, and using nested IF statements.

Formatting Cells

You may have noticed that some data in this tutorial's first spreadsheet was centered in the cells. Here is how to perform that operation:

1. Highlight the cell range to format.
2. Select the Home tab.
3. In the Alignment group, select the Middle Align icon to change the vertical alignment.
4. In the Alignment group, select the Center icon to change the horizontal alignment.

It is also possible to put a border around cells. That treatment might be desirable for highlighting **INPUTS** section cells. To perform that operation:

1. Highlight the cell that needs a border.
2. Select the Home tab.
3. In the Font group, select the drop-down arrow of the Bottom Border icon.
4. Choose the desired border's menu choice; Outside Border and Thick Box Border would be best for your purposes.

You can format numerical values for Currency by following these steps:

1. Highlight the cell or range of cells that should be formatted.
2. Select the Home tab.
3. In the Number group, use the Number Format drop-down arrow to select Currency.

You can format numerical values for decimal places using this procedure:

1. Highlight the cell or range of cells that should be formatted.
2. Select the Home tab.
3. In the Number group, click the Increase Decimal icon once to add one decimal value. Click the Decrease Decimal icon to eliminate a decimal value.

Showing Excel Cell Formulas

If you want to see Excel cell formulas, follow these steps:

1. Press the **Ctrl** key and the back quote key (') at the same time. (The back quote orients from northwest to southeast—on most keyboards, it is to the left of the exclamation point and shares the key with the diacritical tilde mark.)
2. To restore, press the **Ctrl** and (') back quote keys again.

Understanding a Circular Reference

A formula has a circular reference when the reference *refers to the cell the formula is in*. Excel cannot properly evaluate such a formula. The problem is best described by an example. Suppose the formula in cell C18 is =C18–C17. Excel is trying to compute a value for cell C18. To do that, it must evaluate the formula, then put the result on the screen in C18. Excel tries to subtract what is in C17 from what is in C18, but nothing is in C18. Can you see the circularity? To establish a value for C18, Excel must know what is in C18. The process is circular; hence, the term *circular reference*. In the example, the formula in C18 refers to C18. Excel points out circular references, if any exist, when you choose Open for a spreadsheet. Excel also points out circular references as you insert them during the building of a spreadsheet. The software alerts the user to circular references by opening at least one Help window and by drawing arrows between cells involved in the offending formula. You can close the windows, but that will not fix the situation. You must fix the formula that has the circular reference if you want the spreadsheet to give you accurate results.

Using the AND Function and the OR Function in IF Statements

An IF statement has the following syntax:

IF(test condition, result if test is True, result if test is False)

The test conditions in this tutorial's IF statements tested only one cell's value. But a test condition can test more than one value of a cell.

Here is an example from this tutorial's first spreadsheet in which selling price was a function of the economy. Assume, for the sake of illustration, 2009's selling price per unit depends on the economy *and* on the purchase-price outlook. There are two possibilities: (1) If the economic outlook is optimistic *and* the company's purchase-price outlook is down, the selling price will be 1.10 times the prior year's price. (2) In all other cases, the selling price will be 1.03 times the prior year's price. The first possibility's test requires two things to be true *at the same time*: C8 = "O" *AND* C9 = "D." To implement the test, the AND() function is needed. The code in cell C17 would be as follows:

=IF(AND(C8 = "O", C9 = "D"), 1.10 * B17, 1.03 * B17)

When the test that uses the AND() function evaluates to True, the result is 1.10 * B17. When the test that uses the AND() function evaluates to False, the result is the second possibility's outcome: 1.03 * B17.

Now suppose the first possibility is as follows: If the economic outlook is optimistic *or* the purchase-price outlook is down, the selling price will be 1.10 times the prior year's price. Assume in all other cases the selling price will be 1.03 times the prior year's price. Now the test requires *only one of* two things to be true: C8 = "O" *OR* C9 = "D." To implement that, the OR() function would be needed. The code in cell C17 would be:

=IF(OR(C8 = "O", C9 = "D"), 1.10 * B17, 1.03 * B17)

Using IF Statements Inside IF Statements

Recall from the previous section that an IF statement has this syntax:

IF(test condition, result if test is True, result if test is False)

In the examples shown thus far, only two courses of action were possible, so only one test was needed in the IF statement. There can, however, be more than two courses of action; when that is the case, the "result if test is False" clause needs to show further testing. Look at the following example.

Assume again that the 2009 selling price per unit depends on the economy and the purchase-price outlook. Here is the logic: (1) If the economic outlook is optimistic *and* the purchase-price outlook is down, the selling price will be 1.10 times the prior year's price. (2) If the economic outlook is optimistic *and* the purchase-price outlook is up, the selling price will be 1.07 times the prior year's price. (3) In all other cases, the selling price will be 1.03 times the prior year's price. The code in cell C17 would be as follows:

$$=IF(AND(\$C\$8 = ``O", \$C\$9 = ``D"), 1.10 * B17,$$
$$IF(AND(\$C\$8 = ``O", \$C\$9 = ``U"), 1.07 * B17, 1.03 * B17))$$

The first IF statement tests to see if the economic outlook is optimistic and the purchase-price outlook is down. If not, further testing is needed to see whether the economic outlook is optimistic and the purchase-price outlook is up or whether some other situation prevails.

NOTE

The line is broken in the previous example because the page is not wide enough, but in Excel, the formula would appear on one line. The embedded "IF" is not preceded by an equal sign.

Cash Flow Calculations: Borrowings and Repayments

The Scenario Manager cases in this book require you to account for money that the company borrows or repays. Borrowing and repayment calculations are discussed next. At times, you will be asked to think about a question and fill in the answer. Correct responses are found at the end of this section.

To work through the Scenario Manager cases that follow, you must assume two things about a company's borrowing and repayment of debt. First, assume the company wants to have a certain minimum cash level at the end of a year (and thus at the start of the next year). Second, assume a bank will provide a loan to reach the minimum cash level if year-end cash falls short of the desired level.

Here are some numerical examples to test your understanding. Assume NCP stands for "net cash position" and equals beginning-of-the-year cash plus net income after taxes for the year. In other words, the NCP is the cash available at year end, before any borrowing or repayment. For the three examples in Figure C-27, compute the amounts the company needs to borrow to reach its minimum year-end cash level.

Example	NCP	Minimum Cash Required	Amount to Borrow
1	50,000	10,000	?
2	8,000	10,000	?
3	−20,000	10,000	?

FIGURE C-27 Examples of borrowing

One additional assumption you can make is that the company will use its excess cash at year end to pay off as much debt as possible without going below the minimum-cash threshold. Excess cash is the NCP *less* the minimum cash required on hand—amounts over the minimum are available to repay any debt.

In the examples shown in Figure C-28, compute excess cash and then compute the amount to repay. To aid your understanding, you also may want to compute ending cash after repayments.

Example	NCP	Minimum Cash Required	Beginning-of-the-Year Debt	Repay	Ending Cash
1	12,000	10,000	4,000	?	?
2	12,000	10,000	10,000	?	?
3	20,000	10,000	10,000	?	?
4	20,000	10,000	0	?	?
5	60,000	10,000	40,000	?	?
6	−20,000	10,000	10,000	?	?

FIGURE C-28 Examples of repayment

In this section's Scenario Manager cases, your spreadsheet will need two bank financing sections beneath the **INCOME STATEMENT AND CASH FLOW STATEMENT** section: The first section will calculate any needed borrowing or repayment at year's end to compute year-end cash. The second section will calculate the amount of debt owed at the end of the year, after borrowing or repayment of debt.

The first new section, in effect, extends the end-of-year cash calculation, which was shown in Figure C-13. Previously, the amount equaled cash at the beginning of the year plus the year's net income. Now the calculation will include cash obtained by borrowing and cash repaid. Figure C-29 shows the structure of the calculation.

	A	B	C	D
30	NET CASH POSITION (NCP) BEFORE BORROWING AND REPAYMENT OF DEBT (BEGINNING-OF-THE-YEAR CASH PLUS NET INCOME AFTER TAXES)	NA		
31	PLUS: BORROWING FROM BANK	NA		
32	LESS: REPAYMENT TO BANK	NA		
33	EQUALS: END-OF-THE-YEAR CASH ON HAND	10000		

FIGURE C-29   Calculation of END-OF-THE-YEAR CASH ON HAND

The heading in cell A30 was previously END-OF-THE-YEAR CASH ON HAND. But BORROWING increases cash and REPAYMENT OF DEBT decreases cash. So END-OF-THE-YEAR CASH ON HAND is now computed two rows down (in C33 for year 2009 in the example). The value in row 30 must be a subtotal for the BEGINNING-OF-THE-YEAR CASH ON HAND plus the year's NET INCOME AFTER TAXES. That subtotal is called the NET CASH POSITION (NCP) BEFORE BORROWING AND REPAYMENT OF DEBT. (*Note:* Previously, the formula in cell C22 for BEGINNING-OF-THE-YEAR CASH ON HAND was =B30. Now that formula is =B33. It is copied to the right, as before, for the next year.)

The second new section computes end-of-year debt and is called DEBT OWED. That second new section is shown in Figure C-30.

	A	B	C	D
35	DEBT OWED	2008	2009	2010
36	BEGINNING-OF-THE-YEAR DEBT OWED	NA		
37	PLUS: BORROWING FROM BANK	NA		
38	LESS: REPAYMENT TO BANK	NA		
39	EQUALS: END-OF-THE-YEAR DEBT OWED	15000		

FIGURE C-30   DEBT OWED section

As you can see in Figure C-30, at the end of 2008, $15,000 was owed. END-OF-THE-YEAR DEBT OWED equals the BEGINNING-OF-THE-YEAR DEBT OWED plus any new BORROWING FROM BANK (which increases debt owed), less any REPAYMENT TO BANK (which reduces it). So in the example, the formula in cell C39 would be:

=C36+C37−C38

Assume the amounts for BORROWING FROM BANK and REPAYMENT TO BANK are calculated in the first new section. Thus, the formula in cell C37 would be =C31. The formula in cell C38 would be =C32. (BEGINNING-OF-THE-YEAR DEBT OWED is equal to the debt owed at the end of the prior year, of course. The formula in cell C36 for BEGINNING-OF-THE-YEAR DEBT OWED would be an echoed formula. Can you see what it would be? It's an exercise for you to complete. *Hint:* The debt owed at the beginning of a year equals the debt owed at the end of the prior year.)

Now that you have seen how the borrowing and repayment data is shown, the logic of the borrowing and repayment formulas can be discussed.

Calculation of BORROWING FROM BANK

The logic of this in English is:

> If (cash on hand before financing transactions is greater than the
> minimum cash required, then borrowing is not needed;
> otherwise, borrow enough to get to the minimum).

Or (a little more precisely):

> If (NCP is greater than the minimum cash required,
> then BORROWING FROM BANK = 0; otherwise,
> borrow enough to get to the minimum).

Suppose the desired minimum cash at year end is $10,000 and that value is a constant in your spreadsheet's cell C6. Assume the NCP is shown in your spreadsheet's cell C30. The formula (getting closer to Excel) would be as follows:

> IF(NCP > Minimum Cash, 0; otherwise, borrow enough to get to the minimum).

You have cell addresses that stand for NCP (cell C30) and Minimum Cash (C6). To develop the formula for cell C31, substitute the cell addresses for NCP and Minimum Cash. The harder logic is that for the "otherwise" clause. At this point, you should look ahead to the borrowing answers at the end of this section, in Figure C-31. In Example 2, $2,000 was borrowed. Which cell was subtracted from which other cell to calculate that amount? Substitute cell addresses in the Excel formula for 2009's borrowing formula in cell C31:

> =IF(>=   , 0,   -   )

The answer is at the end of this section in Figure C-33.

Calculation of REPAYMENT TO BANK

The logic of this in English is:

> IF(beginning of year debt = 0, repay 0 because nothing is owed, but
> IF(NCP is less than the min, repay zero, because you must *borrow*, but
> IF(extra cash equals or exceeds the debt, repay the whole debt,
> ELSE (to stay above the min, repay only the extra cash))))

Look at the following formula. Assume the repayment will be in cell C32. Assume debt owed at the beginning of the year is in cell C36 and minimum cash is in cell C6. Substitute cell addresses for concepts to complete the formula for 2009 repayment. (Clauses are on different lines because of page width limitations.)

> =IF(= 0, 0,
>   IF(<=   , 0,
>   IF((-   ) >=   ,
>   (-   ))))

The answer is at the end of this section in Figure C-34.

Answers to Questions about Borrowing and Repayment Calculations

Figure C-31 and Figure C-32 give the answers to the questions posed about borrowing and repayment calculations.

Example	NCP	Minimum Cash Required	Borrow	Comments
1	50,000	10,000	Zero	NCP > Min.
2	8,000	10,000	2,000	Need 2,000 to get to Min. 10,000 – 8,000
3	–20,000	10,000	30,000	Need 30,000 to get to Min. 10,000 – (–20,000)

FIGURE C-31   Answers to examples of borrowing

Some notes about the repayment calculations shown in Figure C-32 follow.

Example	NCP	Minimum Cash Required	Beginning-of-the-Year Debt	Repay	Ending Cash
1	12,000	10,000	4,000	2,000	10,000
2	12,000	10,000	10,000	2,000	10,000
3	20,000	10,000	10,000	10,000	10,000
4	20,000	10,000	0	0	20,000
5	60,000	10,000	40,000	40,000	20,000
6	–20,000	10,000	10,000	NA	NA

FIGURE C-32   Answers to examples of repayment

- In Examples 1 and 2, only $2,000 is available for debt repayment (12,000 – 10,000) to avoid going below the minimum cash.
- In Example 3, cash available for repayment is $10,000 (20,000 – 10,000); so all beginning debt can be repaid, leaving the minimum cash.
- In Example 4, no debt is owed, so no debt need be repaid.
- In Example 5, cash available for repayment is $50,000 (60,000 – 10,000); so all beginning debt can be repaid, leaving more than the minimum cash.
- In Example 6, no cash is available for repayment. The company must borrow.

Figure C-33 and Figure C-34 show the calculations for borrowing and repayment of debt.

=IF(C30 >= C6, 0, C6 – C30)

FIGURE C-33   Calculation of borrowing

=IF(C36 = 0, 0, IF(C30 <= C6, 0, IF((C30 – C6) >= C36, C36, (C30 – C6))))

FIGURE C-34   Calculation of repayment

Saving Files after Using Microsoft Excel

As you work, you should make a habit of saving your files periodically. The top of the screen should have a Quick Access Toolbar, and the default Excel installation puts the Save icon there. (The icon looks like a diskette.) You can save your work by clicking that icon. Or you can save by clicking the Office Button, which is the relatively large button in the upper-left corner of the screen. Choose the Save menu choice. The first time you save, you will use the Save As window to specify the following:

- A drive (using the My Documents drop-down menu)
- A filename (using the File name text box)
- A file type (by default, Excel 2007 makes an .xlsx file, but you can save in the Excel 2003 format or in other formats)

You will see the Save As window every time you choose Save As (not Save) when saving.

At the end of your Excel session, save your work using this three-step procedure:

1. Save the file one last time.
2. Using the Office Button, choose Close. If saving to a secondary disk, make sure the disk is still in its drive when you close. Closing the file takes the work off the screen. Do not take your disk out of its drive with the work still on the screen before you close. If you violate that rule, you may lose your work.
3. Using the Office Button, choose Exit Excel. That should take you back to Windows.

CENTRAL'S SPORTS BETTING DECISION

Decision Support Using Excel

PREVIEW

Central is a state in the eastern half of the United States. The state's legislators are considering adopting sports betting as a way of increasing revenues. In this case, you will use the Excel Scenario Manager to model the financial implications of some key variables in the decision.

PREPARATION

Before attempting this case, you should:

- Review spreadsheet concepts discussed in class and/or in your textbook.
- Complete any exercises that your instructor assigns.
- Complete any part of Tutorial C that your instructor assigns, or refer to it as necessary.
- Review file-saving procedures for Windows programs. These are discussed in Tutorial C.
- Refer to Tutorials E and F as necessary.

BACKGROUND

Central is a state in the eastern half of the United States. Central has a gambling program that is run by the state's lottery office. Currently, people can go to any of the state's four horse racing tracks and bet on the races. Also, people can play slot machines and some card games, as if they were at a gambling casino. The race tracks thus are sometimes called "racinos." Gambling has been a significant source of revenue for Central, second only to revenue from personal taxes.

Central is bounded to the north and west by the state of Northwest and to the east by the state of Eastern. These two states have race tracks but do not have other forms of gambling. Therefore, many people in these states who want to gamble will drive to Central to do so.

Northwest and Eastern legislators have been considering allowing "racinos" in their states. The governors of each state have been against legalized slot machine and card game gambling, and so has a narrow majority of the legislators in each state. But these positions could change, especially if tax revenues fall short in these states in the future. If Northwest or Eastern allowed slot machine or card gambling in their states, Central's gambling revenue would go down significantly.

Central state authorities are thinking about expanding their gambling program to collect more revenue. They are considering whether to allow gambling on sports events at new sports gambling lounges within the racinos. Sports bettors could view various sports events beamed in by satellite. Good food and drink would be served so that bettors would have an enjoyable experience, and of course, so that they would stay a while and place multiple bets on sporting events. Sports betting net income would go into the state's general revenue system.

In considering sports betting, Central has an unusual and significant advantage. In 1992 the federal government outlawed sports betting in the United States, except in states that already had a history of legalized sports gambling. Thus, states like Nevada were able to retain their sports gambling program. In the early 1970s, Central tried sports betting for a year but then abandoned the program. The fact that Central once had a program meant that the 1992 law does not prevent Central from trying again. States that border Central are not allowed to adopt a sports betting operation. They could choose to adopt slot machine and card game gambling, but they could not compete with Central's sports betting.

The sports betting operation would be established as a business within Central's Gaming Commission. Significant construction efforts would be needed to expand the racinos. The state would have to borrow an estimated $60 million in construction costs. The $60 million would be the sports gambling operation's debt; eventually it would have to pay off the debt out of its profits. Numerous workers would have to be hired to staff the sports betting operation, and they would have to be properly paid. The operation would have significant administrative expenses, as in any business venture.

The state would have to hire a "sports book agent" to oversee the betting. This agent would set odds on events, record bets, pay winners, collect from losers, and conduct other activities associated with running a sports book. The sports book agent will be expensive, but necessary. Why? To answer that question, think about why, for example, the New York Jets might be favored to beat the Philadelphia Eagles by six points in an NFL game this fall. Why not the other way around? Why not seven points, or five? Who decides? Can any sports fan set the odds? Setting the odds properly requires real expertise, and doing it well is a key profitability factor. (One reason Central's sports book failed in the 1970s is that state politicians thought they could set the odds as well as the experts; they found out they should have stuck with politics.)

Central state officials expect litigation after the operation is up and running. The state's antigambling activists always sue to shut down any change in the scope of legal gambling. The major football, baseball, basketball, and hockey leagues want as few gambling operations as possible in the United States, and have promised to try to shut down a Central sports gambling operation. Central's legal counsel says that the state will be able to defend its operation in court, but that the defense will entail significant legal expenses for a few years.

The state has surveyed current and prospective gamblers. Officials think they have a reasonable estimate of how many people would take advantage of a sports betting operation. There are a couple of significant unknowns, however. Assume that a gambler decides to go to a racino to do some sports betting. How many bets will the average gambler make while he is there? What will be the size of each bet?

The state treasurer says she views sports gambling as a good stand-alone opportunity. But she also sees it as a defensive move intended to support existing slot machine and card game gambling: sports gambling is just one more reason for out-of-state people to come to Central for reasonably priced entertainment. The treasurer says she fears that neighboring states will adopt slot machine and card game gambling, and that this adoption would hurt the sports gambling operation. She says that if Northwest adopts, then Eastern will have to follow suit. She says that the neighboring states would retaliate sooner rather than later, and probably right away.

The state's treasurer has asked you to make a spreadsheet model of the situation. Certain likely scenarios are under consideration. Your spreadsheet will model them in the Scenario Manager.

ASSIGNMENT 1: CREATING A SPREADSHEET FOR DECISION SUPPORT

In this assignment, you will produce a spreadsheet that models the business decision. Then, in Assignment 2, you will write a memorandum to the state treasurer about your analysis and recommendation. In addition, in Assignment 3, you will be asked to prepare an oral presentation of your analysis and recommendation.

First, you will create the spreadsheet model of the proposed program's financial situation. The model is for 10 years of operating sports betting, from 2009 to 2018 inclusive. You will be given guidance on how each spreadsheet section should be set up before entering cell formulas. Your spreadsheet should have the following sections:

- CONSTANTS
- INPUTS
- SUMMARY OF KEY RESULTS
- CALCULATIONS
- INCOME STATEMENT AND CASH FLOW STATEMENT
- DEBT OWED
- AVERAGE RETURN ON INVESTMENT

A discussion of each section follows. *The spreadsheet skeleton is available to you, so you need not type in the skeleton if you do not want to do so.* To access the spreadsheet skeleton, go to your Data files. Select Case 6, then select **SPORTS.xlsx**.

CONSTANTS Section

Your spreadsheet should have the constants shown in Figure 6-1. An explanation of the line items follows the figure. (Years 2014–2018 are not shown here.)

	A	B	C	D	E	F	G
1	**CENTRAL STATE'S SPORTS BETTING PROPOSAL**						
2							
3	**CONSTANTS**	**2008**	**2009**	**2010**	**2011**	**2012**	**2013**
4	SELF-IMPOSED TAX RATE	NA	0.20	0.20	0.20	0.20	0.20
5	CASH NEEDED TO START NEXT YEAR	NA	5,000,000	5,000,000	5,000,000	5,000,000	5,000,000
6	INTEREST RATE ON DEBT	NA	0.06	0.06	0.06	0.06	0.06
7	NUMBER OF POSSIBLE CUSTOMERS	NA	1,000,000	1,005,000	1,010,000	1,015,000	1,020,000
8	EXPECTED ADMINISTRATIVE COSTS	NA	8,000,000	8,080,000	8,160,800	8,242,400	8,324,800
9	COST OF LITIGATION EXPECTED	NA	2,000,000	1,500,000	1,000,000	500,000	0

FIGURE 6-1   CONSTANTS section

- SELF-IMPOSED TAX RATE: The state levies a "tax" on its own money-making operations. Tax receipts are plowed back into the state's social services—for example, taking care of senior citizens, supporting the school systems, building roads, and many other programs. The tax rate on income before taxes from 2009 to 2018 is expected to be 20% each year, as shown.
- CASH NEEDED TO START NEXT YEAR: The state wants its agencies to keep a minimum cash balance. The state says that the sports gambling operation must start each year with a certain amount of cash, as shown.
- INTEREST RATE ON DEBT: The state can borrow money on behalf of its agencies. The state thus becomes the agencies' "bank." The interest rate on state debt is expected to be 6% each year.
- NUMBER OF POSSIBLE CUSTOMERS: Surveys have established how many sports gambling customers there might be each year, as shown. The number shown here represents the number of people who might appear at a racino to gamble on sports in a year. The number increases each year with expected population growth.
- EXPECTED ADMINISTRATIVE COSTS: This is the amount of overhead required to run the sports betting operation. The amount is expected to increase each year.
- COST OF LITIGATION EXPECTED: Outside attorneys will have to be hired to defend the operation in court. The state treasurer says legal problems are expected to go on for four years, with less and less expense.

INPUTS Section

Your spreadsheet should have the inputs shown in Figure 6-2. An explanation of the line items follows the figure. (Years 2014–2018 are not shown here.)

	A	B	C	D	E	F	G
11	**INPUTS**	**2008**	**2009**	**2010**	**2011**	**2012**	**2013**
12	AVERAGE AMOUNT OF SPORTS BET	NA					
13	NEIGHBOR STATES ADOPT GAMBLING?		NA	NA	NA	NA	NA
14	AVERAGE INFLATION RATE IN 10 YEARS		NA	NA	NA	NA	NA
15	AVERAGE NUMBER OF BETS		NA	NA	NA	NA	NA

FIGURE 6-2   INPUTS section

- AVERAGE AMOUNT OF SPORTS BET: This is the amount of each bet. This amount could escalate as the years go on; of course, state officials hope that is the case.
- NEIGHBOR STATES ADOPT GAMBLING?: Will neighboring states adopt gambling in their states? If the spreadsheet user thinks so, the user would enter 'Y'; otherwise, the user would enter 'N'.
- AVERAGE INFLATION RATE IN 10 YEARS: The spreadsheet user enters a value for the average inflation rate in the 10 years from 2009 to 2018. A single value is entered, and is assumed to prevail each year. This rate will affect the expected average salary level for sports gambling workers.
- AVERAGE NUMBER OF BETS: This is the number of bets a gambler places each time he or she comes to a racino. A single value is entered, and is assumed to prevail each year.

Your instructor may tell you to apply Conditional Formatting to the input cells so that out-of-bounds values are highlighted in some way. (For example, the entry could appear in red type or in boldface type). If so, your instructor may provide a tutorial on Conditional Formatting or may ask you to refer to Excel Help.

SUMMARY OF KEY RESULTS Section

Your spreadsheet should contain the results shown in Figure 6-3. An explanation of line items follows the figure. (Years 2014–2018 are not shown here.)

	A	B	C	D	E	F	G
17	SUMMARY OF KEY RESULTS	2008	2009	2010	2011	2012	2013
18	NET INCOME AFTER TAXES	NA					
19	END-OF-THE-YEAR CASH ON HAND	NA					
20	END-OF-THE-YEAR DEBT OWED	NA					
21	AVERAGE RETURN ON INVESTMENT		NA	NA	NA	NA	NA

FIGURE 6-3 SUMMARY OF KEY RESULTS section

For each year, your spreadsheet should show (1) net income after taxes, (2) cash on hand at the end of the year, and (3) the debt owed at the end of the year to bankers, as shown in Figure 6-3. The cells should all be formatted for zero decimals. The spreadsheet should also show the average return on investment for all years. These values are all computed elsewhere in the spreadsheet and should be echoed here. The cells should all be formatted for three decimals.

CALCULATIONS Section

You should calculate various intermediate results that will be used in the income statement and cash flow statement that follows. Calculations, as shown in Figure 6-4, are based on inputs and on year-end 2008 values. When called for, use absolute addressing properly. An explanation of line items follows the figure. (Years 2014–2018 are not shown here.)

	A	B	C	D	E	F	G
23	CALCULATIONS	2008	2009	2010	2011	2012	2013
24	NUMBER OF ACTUAL CUSTOMERS IN YEAR	NA					
25	EXTRA WORKERS NEEDED	NA					
26	AVERAGE YEARLY PAY FOR WORKER	40,000					
27	LABOR EXPENSE IN YEAR	NA					
28	SPORTS BETTING REVENUE	NA					
29	SPORTS BOOK AGENT FEE	NA					
30	INTEREST INCOME	NA					

FIGURE 6-4 CALCULATIONS section

- NUMBER OF ACTUAL CUSTOMERS IN YEAR: If neighboring states are expected to stay out of gambling, this amount is 10% of the number of possible customers in the year. If neighboring states are expected to enter gambling, this amount is only 5% of the number of possible customers in the year. Note that the number of customers is the number of times people show up as sports gambling customers.
- EXTRA WORKERS NEEDED: A worker must be hired for every 1,000 actual customers. Use the Round() function to prevent partial workers (zero decimals).
- AVERAGE YEARLY PAY FOR WORKER: The average is a function of the prior year's average salary and the year's salary change factor, which is the rate of inflation, an input. For example, if the inflation rate is expected to be 1%, 2009's average salary will be 1% greater than 2008's, 2010's average salary will increase or decrease from 2009's, and so on for the remaining years.
- LABOR EXPENSE IN YEAR: This is a function of the number of workers needed in the year and the average yearly pay. Both of these values are calculations.
- SPORTS BETTING REVENUE: This is the amount bet on sports events at the racinos. The amount is a function of the actual number of customers, the average number of bets for each customer, and the average size of a bet.

- SPORTS BOOK AGENT FEE: The sports book agent takes 10% of sports betting revenue. There is no base fee. If betting goes well, the sports book agent does well. If betting is low, the agent does less well.
- INTEREST INCOME: The sports gambling operation will have cash in the "bank" with the state during the year. It will have at least the minimum required, possibly more. The state treasurer will invest this money on a short-term basis and credit the sports betting operation with the interest. The interest rate on short-term cash investments is assumed to be 2% less than the rate on debt owed. The amount of cash invested is the larger of two values: (1) the minimum cash required; (2) the average cash on hand during the year, if greater than the minimum. If the year-end cash were known, the average would be the beginning cash plus the year-end cash divided by two. However, the year-end cash is not known yet, so assume that the average is the cash at the beginning of the year divided by two. To determine the amount invested, you can use the Max() function or an If statement.

INCOME STATEMENT AND CASH FLOW STATEMENT Section

The forecast for net income and cash flow starts with the cash on hand at the beginning of the year. This is followed by the income statement and concludes with the calculation of cash on hand at the year's end. For readability, format cells in this section for zero decimals. Your spreadsheet should look like the ones shown in Figures 6-5 and 6-6. A discussion of line items follows each figure. (Years 2014–2018 are not shown here.)

	A	B	C	D	E	F	G
32	INCOME STATEMENT AND CASH FLOW STATEMENT	2008	2009	2010	2011	2012	2013
33	BEGINNING-OF-THE-YEAR CASH ON HAND	NA					
34							
35	SPORTS BETTING REVENUE	NA					
36	INTEREST INCOME	NA					
37	TOTAL REVENUE	NA					
38	LABOR EXPENSE IN YEAR	NA					
39	SPORTS BOOK AGENT FEE	NA					
40	COST OF LITIGATION	NA					
41	ADMINISTRATIVE COSTS	NA					
42	DEPRECIATION EXPENSE	NA					
43	TOTAL COSTS	NA					
44	INCOME BEFORE INTEREST AND TAXES	NA					
45	INTEREST EXPENSE	NA					
46	INCOME BEFORE TAXES	NA					
47	INCOME TAX EXPENSE	NA					
48	NET INCOME AFTER TAXES	NA					

FIGURE 6-5   INCOME STATEMENT AND CASH FLOW STATEMENT section

- BEGINNING-OF-THE-YEAR CASH ON HAND: This is the cash on hand at the end of the prior year.
- SPORTS BETTING REVENUE: This is a calculated amount that can be echoed here.
- INTEREST INCOME: This is a calculated amount that can be echoed here.
- TOTAL REVENUE: This is the sum of sports betting and interest revenues.
- LABOR EXPENSE IN YEAR: This is a calculated amount that can be echoed here.
- SPORTS BOOK AGENT FEE: This is a calculated amount that can be echoed here.
- COST OF LITIGATION: This is a constant, which can be echoed here.
- ADMINISTRATIVE COSTS: This is a constant, which can be echoed here.
- DEPRECIATION EXPENSE: $60 million is invested in fixed assets. This investment is depreciated over 10 years using the straight-line method, which means the same amount is recorded each year.
- TOTAL COSTS: This is the sum of labor expense, sports book agent fee, litigation costs, administrative costs, and depreciation expense.
- INCOME BEFORE INTEREST AND TAXES: This is the difference between total revenue and total costs.
- INTEREST EXPENSE: This is a function of the debt owed at the *start* of the year and the annual interest rate on debt, a constant.

- INCOME BEFORE TAXES: This is income before interest and taxes, minus interest expense.
- INCOME TAX EXPENSE: This value is zero if the income before taxes is zero or is negative. Otherwise, income tax expense is a function of the year's tax rate and the income before income taxes.
- NET INCOME AFTER TAXES: This is the difference between income before taxes and income tax expense.

Continuing this statement, line items for the year-end cash calculation are discussed. In Figure 6-6, column B is for 2008, column C is for 2009, and so on. Year 2008 values are NA except for END-OF-THE-YEAR CASH ON HAND, which is $5 million. (Years 2014–2018 are not shown here.)

	A	B	C	D	E	F	G
50	NET CASH POSITION (NCP) BEFORE BORROWINGS AND REPAYMENT OF DEBT (BEG OF YEAR CASH + NET INCOME)	NA					
51	ADD: BORROWING FROM STATE	NA					
52	LESS: REPAYMENT TO STATE	NA					
53	EQUALS: END-OF-THE-YEAR CASH ON HAND	5,000,000					

FIGURE 6-6   END-OF-THE-YEAR CASH ON HAND section

- NET CASH POSITION (NCP): The NCP at the end of a year equals cash at the beginning of the year plus the year's net income after taxes.
- ADD: BORROWING FROM STATE: The state is the "bank." Assume that the state will lend the sports betting operation enough money at the end of the year to get to the minimum cash needed to start the next year. If the NCP is less than this minimum, the state must borrow enough money to get to the minimum. Borrowing increases cash on hand, of course.
- LESS: REPAYMENT TO STATE: If the NCP is more than the minimum cash at the end of a year and debt is owed, the sports betting operation must then pay off as much debt as possible (but not take cash below the minimum cash required to start the next year). Repayments reduce cash on hand, of course.
- EQUALS: END-OF-THE-YEAR CASH ON HAND: This equals the NCP plus any borrowings, less any repayments.

DEBT OWED Section

This section shows a calculation of debt owed to the state at year's end, as shown in Figure 6-7. An explanation of line items follows the figure. Year 2008 values are NA, except that the sports operation owes $60 million at the end of that year for the construction funds provided by the state at the start. (Years 2014–2018 are not shown here.)

	A	B	C	D	E	F	G
		2008	2009	2010	2011	2012	2013
55	DEBT OWED						
56	BEGINNING-OF-THE-YEAR DEBT OWED	NA					
57	ADD: BORROWING FROM STATE	NA					
58	LESS: REPAYMENT TO STATE	NA					
59	EQUALS: END-OF-THE-YEAR DEBT OWED	60,000,000					

FIGURE 6-7   DEBT OWED section

- BEGINNING-OF-THE-YEAR DEBT OWED: Debt owed at the beginning of a year equals debt owed at the end of the prior year.
- ADD: BORROWING FROM STATE: This amount has been calculated elsewhere and can be echoed to this section. Borrowings increase the amount of debt owed.
- LESS: REPAYMENT TO STATE: This amount has been calculated elsewhere and can be echoed to this section. Repayments reduce debt owed.
- EQUALS: END-OF-THE-YEAR DEBT OWED: This equals the amount owed at the beginning of a year, plus borrowings in the year, less repayments in the year.

AVERAGE RETURN ON INVESTMENT Section

The spreadsheet body ends with a calculation of the average return on investment, as shown in Figure 6-8. Return on investment is net income in the period divided by the average investment in the period.

	A	B	C	D	E	F	G
61	**AVERAGE RETURN ON INVESTMENT**	**ALL YEARS**					
62	AVERAGE NET INCOME PER YEAR						
63	AVERAGE INVESTMENT IN PERIOD						
64	AVERAGE RETURN ON INVESTMENT						

FIGURE 6-8   AVERAGE RETURN ON INVESTMENT section

- AVERAGE NET INCOME PER YEAR: This is the average net income after taxes in the 10 years. Use the Average() function to compute this.
- AVERAGE INVESTMENT IN PERIOD: The investment starts at $60 million and ends at zero, declining linearly on the straight-line basis.
- AVERAGE RETURN ON INVESTMENT: This equals the average net income divided by the average investment. Note that this amount is echoed to the Summary of Key Results.

ASSIGNMENT 2: USING THE SPREADSHEET FOR DECISION SUPPORT

You will now complete the case by (1) using the spreadsheet to gather the data needed to assess the sports betting idea and (2) documenting your findings and recommendation in a memorandum.

The state treasurer wants to know if the sports betting proposal can make money for the state. The goals are to become profitable and to reduce the state's debt load. "Reduced debt load" means significantly less debt than what the state would start with, $60 million.

The key variables are the average sports bet, the average number of bets per racino visit, the average inflation rate, and whether neighboring states adopt competing gambling programs. The treasurer needs to know what values of these variables will lead to profits and debt reduction.

The treasurer envisions four scenarios that would be reasonable possibilities, and needs to know what the financial results would be in each. In all four scenarios, the average sports bet would be $55 in the first five years and $60 in the last five, and inflation in the period would average 2%. In some scenarios, the average number of bets per visit would be four; in others, the number would be five. In the scenarios, neighboring states either would or would not adopt a competing gambling program.

Assignment 2A: Using the Spreadsheet to Gather Data

You have built the spreadsheet to model the business situation. The treasurer wants to know net income after taxes in 2018, cash on hand at the end of 2018, debt owed at the end of 2018, and the average return on investment in the period in four likely scenarios. The average bet and inflation would be the same in each of the four scenarios. The scenarios are further described in the following shorthand notations:

- *Low roller, Monopoly*: In this scenario, the visitor would bet four times (a "low roller") and the neighboring states would not compete with Central's sports betting program. Central would thus have a monopoly.
- *High roller, Monopoly*: In this scenario, the visitor would bet five times (a "high roller") and the neighboring states would not compete with Central's sports betting program.
- *Low roller, Competition*: In this scenario, the visitor would bet four times and the neighboring states would compete with Central's sports betting program. Central would thus have competition.
- *High roller, Competition*: In this scenario, the visitor would bet five times and the neighboring states would compete with Central's sports betting program.

The financial results in each scenario will act as a negotiating guide to management. You will run "what-if" scenarios with the four sets of input values, using the Scenario Manager. See Tutorial C for the Scenario Manager procedures. Set up the four scenarios. The 13 changing cells are used to input sports bet amounts in each year, neighbor state action, inflation rate, and number of bets. (Note that in the Scenario Manager, you can enter noncontiguous cell ranges as follows: C20..F20, C21, C22. Cell addresses cited here are arbitrary.)

The four Output cells are the 2018 net income, cash on hand, and debt owed cells and the return on investment cell in the Summary of Key Results section. Run the Scenario Manager to gather the data in a report. When you finish, print the spreadsheet with the inputs for any one of the scenarios, and print the Scenario Manager Summary Sheet. Then save the spreadsheet file for the last time (File|Save).

The treasurer hopes that positive net income and significant debt reduction occur in all scenarios. The return on investment should at least cover the cost of debt, 6%. If these events happen, then it would be hard not to undertake the program, even with the threat of competition and litigation.

It may be true that losses and higher debt could appear in one or more scenarios. If so, what variables seem to be most important? Can the program show a profit if other states compete? Can the program make money if the average number of bets is only four? Or, does the visitor need to bet five times?

Assignment 2B: Documenting Your Recommendation in a Memorandum

Open MS Word and write a brief memorandum to the state treasurer, addressing her questions. Tell the treasurer which of the four scenarios would yield 2018 profits and reduced debt, and the related average return on investment. Take a stand: In your opinion, what are the prospects for the sports betting idea? Should the state take a chance on it? Why or why not? If you can see no way forward in the program, state that conclusion. Here is further guidance on your memorandum:

- Your memorandum should be set up as discussed in Tutorial E.
- You need not provide background—the treasurer is aware of the situation. You should briefly state your analytical method and state the results. Give the treasurer your recommendation.
- Support your statements graphically, as your instructor requires: (1) If you used the Scenario Manager, your instructor may want you to go back into Excel and put a copy of the Scenario Manager Summary sheet results into the Windows Clipboard. Then, in Word, copy this graphic into the memorandum. (Tutorial C refers to this procedure.) (2) Or, your instructor may want you to make a summary table in Word, based on the Scenario Manager Summary sheet results, after the first paragraph. The procedure for creating a table in Word is described in Tutorial E.

Your table should resemble the format of the table shown in Figure 6-9.

Scenario	2018 Net Income	2018 Cash on Hand	2018 Debt Owed	Average Return on Investment
Low roller, Monopoly				
High roller, Monopoly				
Low roller, Competition				
High roller, Competition				

FIGURE 6-9   Format of table to insert in memorandum

ASSIGNMENT 3: GIVING AN ORAL PRESENTATION

Your instructor may request that you also present your analysis and recommendation in an oral presentation. If so, assume that the treasurer has accepted your analysis and recommendation. She has asked you to give a presentation explaining your recommendation to the state's senior officials. Prepare to explain your analysis and recommendation to the group in 10 minutes or less. Use visual aids or handouts that you think are appropriate. Tutorial F has guidance on how to prepare and give an oral presentation.

DELIVERABLES

Assemble the following deliverables for your instructor:

1. Printout of your memorandum
2. Spreadsheet printouts
3. Disk, Zip disk, or jump drive, which should have your Word memo file and your Excel spreadsheet file. Do not provide a CD, which would be read-only.

Staple the printouts together, with the memorandum on top. If there is more than one .xlsx file on your disk, write your instructor a note that states the name of your model's .xlsx file.

THE SCHOOL DISTRICT FUNDING DECISION

Decision Support Using Excel

PREVIEW

Your public school district is one of many in your state. The state's Department of Education (DOE) offers to help the district pay for additional teachers. More teachers in the classroom might lead to better student performance, but more teachers mean higher costs. The district must balance its books each year. Should the district accept the state DOE's offer? In this case, you will use the Excel Scenario Manager to model the financial implications of some key variables in the decision.

PREPARATION

Before attempting this case, you should:

- Review spreadsheet concepts discussed in class and/or in your textbook.
- Complete any exercises that your instructor assigns.
- Complete any part of Tutorial C that your instructor assigns, or refer to it as necessary.
- Review file-saving procedures for Windows programs. These are discussed in Tutorial C.
- Refer to Tutorials E and F as necessary.

BACKGROUND

Your state's elected and appointed officials have always prided themselves on the quality of the public education system. In recent years, however, average scores on standardized tests have not improved. In fact, many U.S. government officials think that American public school students should do better on standardized tests. Federal officials have been pressuring state departments of education to improve scores in their states. In fact, federal laws have been enacted that require the states to improve test scores or face the loss of federal education funding.

As you might imagine, thinking about the loss of such funding has concentrated the attention of officials in the states' departments of education. In your own state's DOE, a plan has been proposed that would increase the number of teaching assistants (called "paraprofessionals") in the classroom. More classroom instructors, it is assumed, will lead to more learning and hence better scores on standardized tests.

Before learning the details of the plan, you need to understand how school systems are organized and funded in your state.

The framers of the U.S. Constitution left it to the various states to organize public schooling. Thus, public education has always had a distinctly local flavor in the states. Although federal funding matters very much to the states, the great majority of money spent on schools is raised at the state and local levels.

States organize public education by districts, and each district has its own schools. The district hires the teachers and, working within state DOE guidelines, defines the curriculum. The school district's school board negotiates salary levels with the local teacher's union.

A district is sometimes physically defined by the boundaries of a city. For example, the Philadelphia school district's boundaries are defined by the city limits. Similarly, a district's boundaries might be the boundaries of an entire county. However, most school district boundaries in the United States are set by political bargaining at some point, and then the boundaries rarely change afterward. Some states have very few school districts, while other states have many districts.

A local school district gets money from the state government and from local taxes. State-level funding comes from a state sales tax or from revenue generated by the state's income tax. Local funding is usually raised by taxing real estate. In some states, funding from the state government is very high, and so local taxes can be low. In other states, there is a rough balance or the reverse is true.

Your state's government provides more than half of the funding to the local districts. The rest of the money is raised by local property taxes, following a tax rate that the school board sets each year.

For every 12 students, state funds are provided to pay part of a teacher's salary. (Currently, this amount is $40,000.) The local school district works with the local teacher's union to set the salary level in the district. Currently, the average salary level in your local district is $52,000. The district's local taxes must make up for the $12,000 that is not paid by the state.

Of course, a school district has expenses other than teacher salaries. For example, buses must be hired to transport students, books must be bought, buildings heated, sports teams funded, and so forth. In your state, a district must pay these expenses out of local property taxes.

In your state, a district must at least break even each year. In other words, a district is not allowed to spend more money than it has. In extreme cases, a district that runs a deficit is bailed out by the state DOE, which will make a short-term loan to the district. Such situations are embarrassing to the local school board and local school districts, and they are hated at the state level because the state must borrow money in the bond market to make the loan. Thus, a school district's officials and local school board work hard to budget their money during the school year.

Oddly, the reverse situation—spending much less than what is available—is not desirable, either. A district that has built huge cash reserves looks like it has too high a local tax rate, or that it could be spending more on educating children. In fact, your state's rules require a district to maintain a cash reserve roughly equal to two month's worth of teacher salaries. Your district tries to maintain that much money in reserve, but not more.

In fact, a few years ago, your district's finance office was not good at budgeting. The district wound up being obligated to pay for ambitious education programs for which it did not have money. The state loaned the district money to cover its deficit, and the district has been paying off the loan slowly but surely. Currently, the loan balance is $2 million.

In all the states, students are tested frequently for knowledge in reading, mathematics, and in some other courses. These tests are "standardized"—that is, every student in a grade takes the same test, and the grading is standardized as well. Students are tested using exams that are developed by the state and at the federal level. The underlying assumptions are that all this testing measures what has been learned, and that if students can be pushed to do better on tests, they must be better educated than they were previously.

Your state's DOE has proposed to fund more teaching assistants to help regular classroom teachers. A teaching assistant would be called a "paraprofessional," or "para." Paras would specialize in topics such as reading or mathematics and would work with students who should be doing better than they are. The general idea is that more teachers per student in the schools will mean more learning, which will mean better test scores. Better test scores mean that the federal government will not withdraw education funding from the state.

The state has proposed a funding formula for paras. It is up to each district to decide if it wants to participate in the paraprofessional program. Furthermore, the districts that participate must decide how many paras they will hire. Of course, any district would want to have more teachers—education research has shown a correlation between staffing and learning. But your district wants to be sure that the plan will not result in another deficit situation!

The following sections provide further details on how the district's tax rate is established and on how the state's paraprofessional funding would work.

District Tax Rate

Your state has laws that govern the property tax rate computation. The rate is equal to the following ratio:

$$(\text{Locally funded costs}/\text{Current year tax base}) + .003$$

"Locally funded costs" equal total expenses minus cash provided by the state. For example, say that total district costs are $90 million and that the state will provide $50 million for teacher salaries. Locally funded costs are $40 million. The tax base is the assessed value of the real estate owned by people within the geographic boundaries of the district. Real estate owned includes single-family homes, homes and apartments for rent, and commercial buildings. An assessed value is established by local and county officials, and roughly corresponds to the real estate's approximate market value. Continuing the example, if the tax base is $500 million in the district, the tax rate for the year would be:

$$(\$40 \text{ million}/\$500 \text{ million}) + .003 = .083$$

Your district has two other rules for computing the tax rate:

- In the current year, actual costs are not known until after the year is over, but a rate is needed at the start of the year. So, to compute a tax rate for year X, locally funded costs from year X–1 are used. This means that year X–1 values would be used for total expenses and for amounts contributed by the state.
- The tax rate in year X cannot exceed the rate in year X–1 by more than .01. For example, the rate in year X–1 might be .062. In year X, the formula might yield a rate of .073. That rate would be too high, so a rate of .072 would be used in year X.

State Paraprofessional Funding Plan

The state offers to pay $10,000 toward a para's salary in 2009 as long as the district will pay that much as well. In future years the state offers to increase this amount by one half the planned growth rate in regular teacher salary. The district would be expected to give para pay raises in the same way.

In addition, if paras are hired, the state will send extra "excellence funds" to the district, based on the assumption that test scores will improve with paras in the classroom.

The test score increment is the test score improvement expected to occur if paras are hired. The increment is computed by the following formula:

$$(12 - (\text{the ratio of students to teachers and paras}))/100$$

The factor 12 is the current teacher-student funding factor, and is used as a baseline here. The ratio of students to teachers and paras is the total number of students divided by the sum of teachers and paraprofessionals.

In setting this formula, state DOE officials realize that odd results could occur if many paraprofessionals are hired. Thus, there is a limit to the test score increment that the DOE will honor. In practice, average scores on standardized tests in a district do not increase very much year to year. Thus, the highest increment that the state will allow a district to use is .02. For example, a district might compute an expected test score increment of (12–9)/100 = .03. Because the maximum is .02, however, the state will only allow .02 to be used in computing excellence funds for the district. If a district computes an expected test score increment of (12–11)/100 = .01, the state would require .01 to be used in computing excellence funds for the district.

Excellence funds per para would be calculated by the following formula:

$$(\$7,500 * \text{test score increment} * 100)$$

Thus, if the test score increment was calculated as .01, the state would provide $7,500 per para hired by the district.

The school district's chief financial officer (CFO) says everyone in education thinks that having more teachers in the classroom is a good thing. But the CFO also says that the district must be able to afford the paras. The CFO says that the district is currently doing well financially, and could opt out of the para program and try to improve test scores without the paras. The CFO says that student growth is expected in the next five years, but that the rate of growth is not clear. The CFO says the district would want to avoid a plan that looked financially acceptable at one level of student growth, but that put the district in a deficit at another level of growth.

The CFO has asked you to make a spreadsheet model of the situation. Certain likely scenarios are under consideration. Your spreadsheet will model them in the Scenario Manager.

ASSIGNMENT 1: CREATING A SPREADSHEET FOR DECISION SUPPORT

In this assignment, you will produce a spreadsheet that models the business decision. Then, in Assignment 2, you will write a memorandum to the district's CFO about your analysis and recommendation. In addition, in Assignment 3, you will be asked to prepare an oral presentation of your analysis and recommendation.

First, you will create the spreadsheet model of the proposed program's financial situation. The model is for the next five years, from 2009 to 2013 inclusive. You will be given guidance on how each spreadsheet section should be set up before entering cell formulas. Your spreadsheet should have the following sections:

- CONSTANTS
- INPUTS
- SUMMARY OF KEY RESULTS
- CALCULATIONS
- INCOME STATEMENT AND CASH FLOW STATEMENT
- DEBT OWED

A discussion of each section follows. *The spreadsheet skeleton is available to you, so you need not type in the skeleton if you do not want to do so.* To access the spreadsheet skeleton, go to your Data files. Select Case 7, then select **SCHOOL.xlsx**.

CONSTANTS Section

Your spreadsheet should have the constants shown in Figure 7-1. An explanation of the line items follows the figure.

	A	B	C	D	E	F	G
1	**THE SCHOOL DISTRICT FUNDING DECISION**						
2							
3	CONSTANTS	2008	2009	2010	2011	2012	2013
4	TAX RATE	NA	0.20	0.20	0.20	0.20	0.20
5	TEACHER UNIT FACTOR	NA	12	12	12	12	12
6	TAX BASE	NA	500,000,000	515,000,000	530,000,000	550,000,000	570,000,000
7	GROWTH RATE -- DISTRICT SALARY	NA	0.02	0.03	0.04	0.04	0.04
8	GROWTH RATE -- STATE TEACHER SALARY	NA	0.03	0.03	0.03	0.03	0.03
9	ADMINISTRATIVE EXPENSES	NA	10,000,000	10,100,000	10,200,000	10,300,000	10,400,000
10	CASH NEEDED TO START NEXT YEAR	NA	3,500,000	3,500,000	3,500,000	3,500,000	3,500,000
11	INTEREST RATE ON DEBT OWED	NA	0.06	0.06	0.06	0.06	0.06

FIGURE 7-1 CONSTANTS section

- TAX RATE: The state levies a "tax" on profits that districts generate in a school year. The money collected is set aside in a special projects fund that the state DOE can use as it sees fit. The tax rate on income before taxes from 2009 to 2013 is expected to be 20% each year, as shown.
- TEACHER UNIT FACTOR: The state will fund a teacher for every 12 students in a district.
- TAX BASE: The assessed valuation of real estate in the district is expected to be $500 million in 2009, increasing slightly each year as shown.
- GROWTH RATE – DISTRICT SALARY: By union agreement the district is obligated to give a pay raise each year, as shown. For example, in 2010, the district must pay 3% more than in 2009.
- GROWTH RATE – STATE TEACHER SALARY: The state has a policy of increasing its average teacher salary contribution by 3% a year.
- ADMINISTRATIVE EXPENSES: Nonsalary district expenses are expected to be $10 million in 2009, increasing each year as shown.
- CASH NEEDED TO START NEXT YEAR: The state requires the districts to have a minimum cash balance at the start of a school year, as shown.
- INTEREST RATE ON DEBT OWED: The state can borrow money on behalf of districts that get into financial trouble. The state thus becomes the district's "bank." The interest rate on state debt is expected to be 6% each year.

INPUTS Section

Your spreadsheet should have the inputs shown in Figure 7-2. An explanation of the line items follows the figure.

	A	B	C	D	E	F	G
13	INPUTS	2008	2009	2010	2011	2012	2013
14	STUDENT GROWTH	NA		NA	NA	NA	NA
15	NUMBER OF TEACHERS PER PARA	NA					

FIGURE 7-2   INPUTS section

- STUDENT GROWTH: How will the district budget look with different numbers of students? The user enters a decimal value here to indicate how much the total number of students will grow each year in the five years. For example, .02 would imply a 2% growth rate each year. The number of students in the district in 2009 will be 2% more than in 2008, the number of students in 2010 will be 2% more than in 2009, and so on.
- NUMBER OF TEACHERS PER PARA: It is up to the district to decide how many paras it will hire in the program. Here, the user enters a number indicating the teacher-para ratio. For example, if a 1 is entered, there will be one para for each teacher; in other words, each teacher would have an assistant teacher! If the number entered is 12, then a para would be hired for every 12 teachers. If a 0 is entered, the district has opted to not participate in the program—no paras will be hired. Note that the ratio could be a different value each year.

Your instructor may tell you to apply Conditional Formatting to the input cells so that out-of-bounds values are highlighted in some way. (For example, the entry could appear in red type or in boldface type.) If so, your instructor may provide a tutorial on Conditional Formatting or may ask you to refer to Excel Help.

SUMMARY OF KEY RESULTS Section

Your spreadsheet should contain the results shown in Figure 7-3. An explanation of line items follows the figure.

	A	B	C	D	E	F	G
17	SUMMARY OF KEY RESULTS	2008	2009	2010	2011	2012	2013
18	NET INCOME AFTER TAXES	NA					
19	END-OF-THE-YEAR CASH ON HAND	NA					
20	END-OF-THE-YEAR DEBT OWED	NA					

FIGURE 7-3   SUMMARY OF KEY RESULTS section

For each year, your spreadsheet should show (1) net income after taxes, (2) cash on hand at the end of the year, and (3) the debt owed at the end of the year to the state, as shown in Figure 7-3. The cells should all be formatted for zero decimals.

CALCULATIONS Section

You should calculate various intermediate results that will be used in the income statement and cash flow statement that follows. Calculations, as shown in Figure 7-4, are based on inputs and on year-end 2008 values. When called for, use absolute addressing properly. An explanation of line items follows the figure.

	A	B	C	D	E	F	G
22	**CALCULATIONS**	**2008**	**2009**	**2010**	**2011**	**2012**	**2013**
23	NUMBER OF STUDENTS IN DISTRICT	20,000					
24	NUMBER OF TEACHERS FUNDED BY STATE	NA					
25	TEACHER SALARY -- STATE	40,000					
26	TEACHER SALARY -- DISTRICT	12,000					
27	TOTAL SALARY EXPENSE	NA					
28	TAX RATE FOR CURRENT YEAR	NA	0.062				
29	NUMBER OF PARA-PROFESSIONALS	NA					
30	RATIO OF STUDENTS TO TEACHERS+PARAS	NA					
31	PARA SALARY -- STATE	NA	10,000				
32	PARA SALARY -- DISTRICT	NA	10,000				
33	AVERAGE TEST SCORE INCREMENT	NA					
34	AVERAGE STATE TEST SCORE	0.6000					
35	EXCELLENCE FUNDS PER PARA	NA					

FIGURE 7-4   CALCULATIONS section

- NUMBER OF STUDENTS IN DISTRICT: The number of students in the district in a year is a function of the number in the prior year, and the expected student growth rate, an input value. If the growth rate is expected to be 1%, 2009's number will be 101% of 2008's number, 2010's number will be 101% of 2009's, and so on. Use the Round() function to prevent partial students (zero decimals).
- NUMBER OF TEACHERS FUNDED BY STATE: This is a function of the number of students in the district and the teacher unit factor, a constant value. Use the Round() function to prevent partial teachers (zero decimals).
- TEACHER SALARY – STATE: The average amount that the state will contribute toward a teacher's salary is a function of the prior year's average, and the growth rate expected in state salary contribution, a constant value. For example, in 2008 the average state contribution was $40,000. This amount is expected to go up by 3% in 2009.
- TEACHER SALARY – DISTRICT: The average amount that the district will contribute toward a teacher's salary is a function of the prior year's average, and the growth rate expected in district salary contribution, a constant value. For example, in 2008 the average district contribution was $12,000. This amount is expected to go up by 2% in 2009.
- TOTAL SALARY EXPENSE: This is a function of the number of teachers needed during the year and the total average yearly salary. The total average salary is the sum of the state and district contributions.
- TAX RATE FOR CURRENT YEAR: This calculation is described in detail in the Background section. Keep in mind that prior year district expenses and amounts contributed by the state are used in the calculation. Note also that amounts contributed by the state can be used for teacher and para salaries and for excellence funds. Also keep in mind that the rate cannot go up more than .01 from year to year—consider using the Min() function to enforce that requirement. The tax rate for 2009 has already been computed and can be entered as .062. A formula should be entered to compute the rate for 2010 through 2013.
- NUMBER OF PARAPROFESSIONALS: This is a function of the number of teachers needed in a year, a calculation, and of the number of teachers per para factor, an input. If the teachers per para input is a zero, then no paras will be hired.
- RATIO OF STUDENTS TO TEACHERS + PARAS: This is the ratio of the students in the district to the sum of teachers and paras. Use the Round() function to prevent partial values (zero decimals).
- PARA SALARY – STATE: This is the amount that the state would contribute to a para's salary. $10,000 is the amount for 2009, so that amount can be entered. A formula should be entered to compute the rate for 2010 through 2013. In those years the contribution is a function of the prior year's value, increased by one half the expected growth in the state teacher salary contribution, a constant.
- PARA SALARY – DISTRICT: This is the amount that the district would contribute to a para's salary. $10,000 is the amount for 2009, so that amount can be entered. A formula should be entered to compute the rate for 2010 through 2013. In those years the contribution is a function of the prior year's value, increased by one half the expected growth in the district teacher salary contribution, a constant.

- AVERAGE TEST SCORE INCREMENT: This calculation is described in detail in the Background section: the difference between 12 and the year's ratio of students to teachers and paras, divided by 100. However, the state will not honor an increment greater than .02. Consider using the MIN() function in this calculation.
- AVERAGE STATE TEST SCORE: This year's average test score is a function of last year's score plus this year's computed test score increment.
- EXCELLENCE FUNDS PER PARA: This amount is $7,500 times the year's test score increment times 100.

INCOME STATEMENT AND CASH FLOW STATEMENT Section

The forecast for net income and cash flow starts with the cash on hand at the beginning of the year. This is followed by the income statement and concludes with the calculation of cash on hand at the year's end. For readability, format cells in this section for zero decimals. Your spreadsheet should look like the ones shown in Figures 7-5 and 7-6. A discussion of line items follows each figure.

	A	B	C	D	E	F	G
37	INCOME STATEMENT AND CASH FLOW STATEMENT	2008	2009	2010	2011	2012	2013
38	BEGINNING-OF-THE-YEAR CASH ON HAND	NA					
39							
40	REVENUES:	NA					
41	TEACHER SALARY -- STATE	NA					
42	PARA PROFESSIONAL SALARY -- STATE	NA					
43	PROPERTY TAXES	NA					
44	EXCELLENCE FUNDS	NA					
45	TOTAL REVENUE	NA					
46	COSTS:	NA					
47	TEACHERS SALARY	NA					
48	PARA-PROFESSIONAL SALARY	NA					
49	ADMINISTRATIVE EXPENSES	NA					
50	TOTAL COSTS	NA					
51	INCOME BEFORE INTEREST AND TAXES	NA					
52	INTEREST EXPENSE	NA					
53	INCOME BEFORE TAXES	NA					
54	INCOME TAX EXPENSE	NA					
55	NET INCOME AFTER TAXES	NA					

FIGURE 7-5   INCOME STATEMENT AND CASH FLOW STATEMENT section

- BEGINNING-OF-THE-YEAR CASH ON HAND: This is the cash on hand at the end of the prior year.
- TEACHER SALARY – STATE: This is a function of the number of teachers and the state's contribution toward their salary.
- PARAPROFESSIONAL SALARY – STATE: This is a function of the number of paraprofessionals and the state's contribution toward their salary.
- PROPERTY TAXES: This is a function of the tax rate for the year (a calculation) and the current tax base (a constant).
- EXCELLENCE FUNDS: This is a function of the number of paraprofessionals and the excellence funds received per para.
- TOTAL REVENUE: This is the sum of all four forms of revenue.
- TEACHERS SALARY: This is total teacher salary expense, a calculation.
- PARAPROFESSIONAL SALARY: This is a function of the number of paraprofessionals and the sum of their state and district salaries.
- ADMINISTRATIVE EXPENSES: This is a constant, which can be echoed here.
- TOTAL COSTS: This is the sum of teacher and paraprofessional salaries and administrative expenses.
- INCOME BEFORE INTEREST AND TAXES: This is the difference between total revenue and total costs.
- INTEREST EXPENSE: This is a function of the debt owed to the state at the *start* of the year and the interest rate on debt, a constant.
- INCOME BEFORE TAXES: This is income before interest and taxes, minus interest expense.

- INCOME TAX EXPENSE: This value is zero if the income before taxes is zero or is negative. Otherwise, income tax expense is a function of the year's tax rate and the income before income taxes. The tax rate is a constant.
- NET INCOME AFTER TAXES: This is the difference between income before taxes and income tax expense.

Continuing this statement, line items for the year-end cash calculation are discussed. In Figure 7-6, column B is for 2008, column C is for 2009, and so on. Year 2008 values are NA except for END-OF-THE-YEAR CASH ON HAND, which is $3.5 million.

	A	B	C	D	E	F	G
57	NET CASH POSITION (NCP) BEFORE BORROWING AND REPAYMENT OF DEBT (BEG OF YR CASH + NET INCOME)	NA					
58	ADD:BORROWING FROM STATE	NA					
59	LESS:REPAYMENT TO STATE	NA					
60	EQUALS: END-OF-THE-YEAR CASH ON HAND	3,500,000					

FIGURE 7-6 END-OF-THE-YEAR CASH ON HAND section

- NET CASH POSITION (NCP): The NCP at the end of a year equals cash at the beginning of the year plus the year's net income after taxes.
- ADD: BORROWING FROM STATE: The state is the "bank." Assume that the state will lend the district enough money at year's end to get to the minimum cash needed to start the next year. If the NCP is less than this minimum, the district must borrow enough money to get to the minimum. Borrowing increases cash on hand, of course.
- LESS: REPAYMENT TO STATE: If the NCP is more than the minimum cash at the end of a year and debt is owed, the district must then pay off as much debt as possible (but not take cash below the minimum cash required to start the next year). Repayments reduce cash on hand, of course.
- EQUALS: END-OF-THE-YEAR CASH ON HAND: This amount equals the NCP plus any borrowings, less any repayments.

DEBT OWED Section

This section shows a calculation of debt owed to the state at year's end, as shown in Figure 7-7. An explanation of line items follows the figure. Year 2008 values are NA, except that the district owes $2 million at the end of that year for the balance of a loan previously made by the state to the district.

	A	B	C	D	E	F	G
62	DEBT OWED	2008	2009	2010	2011	2012	2013
63	BEGINNING-OF-THE-YEAR DEBT OWED	NA					
64	ADD:BORROWING FROM STATE	NA					
65	LESS:REPAYMENT TO STATE	NA					
66	EQUALS: END-OF-THE-YEAR DEBT OWED	2,000,000					

FIGURE 7-7 DEBT OWED section

- BEGINNING-OF-THE-YEAR DEBT OWED: The debt owed to the state at the beginning of a year equals debt owed at the end of the prior year.
- ADD: BORROWING FROM STATE: This amount has been calculated elsewhere and can be echoed to this section. Borrowings increase the amount of debt owed.
- LESS: REPAYMENT TO STATE: This amount has been calculated elsewhere and can be echoed to this section. Repayments reduce the amount of debt owed.
- EQUALS: END-OF-THE-YEAR DEBT OWED: This amount equals the amount owed at the beginning of a year, plus borrowings in the year, less repayments in the year.

ASSIGNMENT 2: USING THE SPREADSHEET FOR DECISION SUPPORT

You will now complete the case by (1) using the spreadsheet to gather the data needed to assess the paraprofessional opportunity and (2) documenting your findings and recommendation in a memorandum.

The CFO says that district officials would want to use paraprofessionals in the classroom because it should promote student achievement. However, the CFO says that district officials also have financial goals. State debt must decrease over time. This means that the district must run a surplus and increase cash on hand, if possible. With this goal in mind, the CFO says that district officials have many specific questions:

124

- Ideally, district officials would like to see high student growth and as many paras as possible—what manager does not want his or her kingdom to expand? The CFO wants to know the financial results if there is high growth in students and as many paras as teachers.
- At the other end of the spectrum, some district officials are happy the way things are now: no student growth and no paras. The CFO wants to know the financial results if there is no growth in students and no paras.
- Most district officials are conservative in the following sense: If paras are hired, financial goals should be met no matter what student growth is (none, 1%, or 2%). The CFO wants to know if the state's para program is "balanced"—will the district meet financial goals regardless of student growth? Another way of looking at this question is: For what student growth levels (if any) should the district *not* hire paras?
- Some district officials want to know the answer to the following question: For each growth level, what is the financially best para hiring strategy (none, 1:1, 6:1, 12:1)?
- Some district officials want to know the answer to the following questions: Under what combination of student growth and para hiring does the district have the best financial results? Under what combination of student growth and para hiring does the district have the worst financial results?
- Finally, if the para hiring program does not seem financially beneficial, why is it not? The state tries to provide help with para salaries and excellence funds—is that not good enough?

The key variables are the expected growth rate in students and the teacher-para factor. Expected growth rate can take on values of 0, .01, and .02. The teacher-para factor can take on the four values of 0, 1, 6, and 12. Thus, the CFO envisions 12 scenarios that would be reasonable possibilities, and needs to know the financial results in each scenario.

Assignment 2A: Using the Spreadsheet to Gather Data

You have built the spreadsheet to model the business situation. The CFO wants to know net income after taxes in 2013, cash on hand at the end of 2013, and debt owed at the end of 2013 in the 12 likely scenarios. The scenarios are further described in the following shorthand notations:

- **Low0**: In this scenario, student growth is zero ("Low") and no paras are hired.
- **Low1**: In this scenario, student growth is zero and a para is hired for every teacher.
- **Low6**: In this scenario, student growth is zero and a para is hired for every 6 teachers.
- **Low12**: In this scenario, student growth is zero and a para is hired for every 12 teachers.
- **Mid0**: In this scenario, student growth is 1% ("Mid") and no paras are hired.
- **Mid1**: In this scenario, student growth is 1% and a para is hired for every teacher.
- **Mid6**: In this scenario, student growth is 1% and a para is hired for every 6 teachers.
- **Mid12**: In this scenario, student growth is 1% and a para is hired for every 12 teachers.
- **High0**: In this scenario, student growth is 2% ("High") and no paras are hired.
- **High1**: In this scenario, student growth is 2% and a para is hired for every teacher.
- **High6**: In this scenario, student growth is 2% and a para is hired for every 6 teachers.
- **High12**: In this scenario, student growth is 2% and a para is hired for every 12 teachers.

The financial results in each scenario will act as a guide to answering district management's questions. These answers will help management decide if it should support the paraprofessional program. You will run "what-if" scenarios with the 12 sets of input values, using the Scenario Manager. See Tutorial C for the Scenario Manager procedures. Set up the 12 scenarios. The six changing cells are used to input student growth and the teacher-para factors. (Note that in the Scenario Manager, you can enter noncontiguous cell ranges as follows: C20..F20, C21, C22. Cell addresses cited here are arbitrary.)

The three Output cells are the 2013 net income, cash on hand, and debt owed cells in the Summary of Key Results section. Run the Scenario Manager to gather the data in a report. When you finish, print the spreadsheet with the inputs for any one of the scenarios, and print the Scenario Manager Summary Sheet. Then save the spreadsheet file for the last time (File|Save).

Assignment 2B: Documenting Your Recommendation in a Memorandum

Open MS Word and write a brief memorandum to the CFO, addressing his questions. In your memorandum, answer the CFO's questions based on the data generated using the Scenario Manager. Take a stand: In your opinion, what are the prospects for the paraprofessional program? Should the district join in? Why or why not? Here is further guidance on your memorandum:

- Your memorandum should be set up as discussed in Tutorial E.
- You need not provide background—the CFO is aware of your work. You should briefly state your analytical method and state the results, in the form of answering the CFO's questions. Then give the CFO your recommendation.
- Support your statements graphically, as your instructor requires: (1) If you used the Scenario Manager, your instructor may want you to go back into Excel and put a copy of the Scenario Manager Summary sheet results into the Windows Clipboard. Then, in Word, copy this graphic into the memorandum. (Tutorial C refers to this procedure.) (2) Or, your instructor may want you to make a summary table in Word, based on the Scenario Manager Summary sheet results, after the first paragraph. The procedure for creating a table in Word is described in Tutorial E.

Your table should resemble the format of the table shown in Figure 7-8.

Scenario	2013 Net Income	2013 Cash on Hand	2013 Debt Owed
Low0			
Low1			
Low6			
Low12			
Mid0			
Mid1			
Mid6			
Mid12			
High0			
High1			
High6			
High12			

FIGURE 7-8 Format of table to insert in memorandum

ASSIGNMENT 3: GIVING AN ORAL PRESENTATION

Your instructor may request that you also present your analysis and recommendation in an oral presentation. If so, assume that the CFO has accepted your analysis and recommendation, and has asked you to give a presentation explaining your recommendation to the district's management. Prepare to explain your analysis and recommendation to the group in 10 minutes or less. Use visual aids or handouts that you think are appropriate. Tutorial F has guidance on how to prepare and give an oral presentation.

DELIVERABLES

Assemble the following deliverables for your instructor:

1. Printout of your memorandum
2. Spreadsheet printouts
3. Disk, Zip disk, or jump drive, which should have your Word memo file and your Excel spreadsheet file. Do not provide a CD, which would be read-only.

Staple the printouts together, with the memorandum on top. If there is more than one .xlsx file on your disk, write your instructor a note that states the name of your model's .xlsx file.

PART 3

DECISION SUPPORT CASES
USING THE EXCEL SOLVER

TUTORIAL **D**

BUILDING A DECISION SUPPORT SYSTEM USING THE EXCEL SOLVER

Decision Support Systems (DSS) help people make decisions. (The nature of DSS programs is discussed in Tutorial C.) Tutorial D teaches you how to use the Solver, one of the Excel built-in decision support tools.

For some business problems, decision makers want to know the best, or optimal, solution. Usually that means maximizing a variable (for example, net income) or minimizing another variable (for example, total costs). This optimization is subject to constraints, which are rules that must be observed when solving a problem. The Solver computes answers to such optimization problems.

This tutorial has four sections:

1. **Using the Excel Solver**: In this section, you'll learn how to use the Solver in decision making. As an example, you use the Solver to create a production schedule for a sporting goods company. This schedule is called the Base Case.
2. **Extending the Example**: In this section, you'll test what you've learned about using the Solver as you modify the sporting goods company's production schedule. This is called the Extension Case.
3. **Using the Solver on a New Problem**: In this section, you'll use the Solver on a new problem.
4. **Troubleshooting the Solver**: In this section, you'll learn how to overcome problems you might encounter when using the Solver.

NOTE

Tutorial C offers some guidance on basic Excel concepts, such as formatting cells and using functions such as =IF(). Refer to Tutorial C for a review of those topics.

USING THE EXCEL SOLVER

Suppose a company must set a production schedule for its various products, each of which has a different profit margin (selling price, less costs). At first, you might assume the company will maximize production of all profitable products to maximize net income. However, a company typically cannot make and sell an unlimited number of its products because of constraints.

One constraint affecting production is the shared resource problem. For example, several products in a manufacturer's line might require the same raw materials, which are in limited supply. Similarly, the manufacturer might require the same machines to make several of its products. In addition, there also might be a limited pool of skilled workers available to make the products.

In addition to production constraints, sometimes management's policies impose constraints. For example, management might decide that the company must have a broader product line. As a consequence, a certain production quota for several products must be met, regardless of profit margins.

Thus, management must find a production schedule that will maximize profit given the constraints. Optimization programs like the Solver look at each combination of products, one after the other, ranking each combination by profitability. Then the program reports the most profitable combination.

To use the Solver, you'll set up a model of the problem, including the factors that can vary, the constraints on how much they can vary, and the goal—that is, the value you are trying to maximize (usually net income) or minimize (usually total costs). The Solver then computes the best solution.

Setting Up a Spreadsheet Skeleton

Suppose your company makes two sporting goods products—basketballs and footballs. Assume you will sell all of the balls you produce. To maximize net income, you want to know how many of each kind of ball to make in the coming year.

Making each kind of ball requires a certain (and different) number of hours, and each ball has a different raw materials cost. Because you have only a limited number of workers and machines, you can devote a maximum of 40,000 hours to production, which is a shared resource. You do not want that resource to be idle, however. Downtime should be no more than 1,000 hours in a year, so machines should be used for at least 39,000 hours.

Marketing executives say that you cannot make more than 60,000 basketballs and cannot make fewer than 30,000. Furthermore, they say that you must make at least 20,000 footballs but not more than 40,000. Marketing also says that the ratio of basketballs to footballs produced should be between 1.5 and 1.7—that is, more basketballs than footballs, but within limits.

What would be the best production plan? This problem has been set up in the Solver. The spreadsheet sections are discussed in the pages that follow.

AT THE KEYBOARD

Start by saving the blank spreadsheet as **SPORTS1.xlsx**. Then enter the skeleton and formulas as they are discussed.

CHANGING CELLS Section

The **CHANGING CELLS** section contains the variables the Solver is allowed to change while it looks for the solution to the problem. Figure D-1 shows the skeleton of this spreadsheet section and the values you should enter. An analysis of the line items follows the figure.

	A	B	C	D	E
1	**SPORTING GOODS EXAMPLE**				
2	CHANGING CELLS				
3	NUMBER OF BASKETBALLS	1			
4	NUMBER OF FOOTBALLS	1			

FIGURE D-1   CHANGING CELLS section

- The changing cells are for the number of basketballs and footballs to be made and sold. The changing cells are like input cells, except the Solver (not you) plays "what if" with the values, trying to maximize or minimize some value (in this case, maximize net income).
- Some number should be put in the changing cells each time before the Solver is run. It is customary to put the number 1 in the changing cells (as shown). The Solver will change these values when the program is run.

CONSTANTS Section

Your spreadsheet also should have a section for values that will not change. Figure D-2 shows a skeleton of the **CONSTANTS** section and the values you should enter. A discussion of the line items follows the figure.

NOTE

You should format cells in the constants range to two decimal places.

	A	B	C	D	E
6	**CONSTANTS**				
7	BASKETBALL SELLING PRICE	14.00			
8	FOOTBALL SELLING PRICE	11.00			
9	TAX RATE	0.28			
10	NUMBER OF HOURS TO MAKE A BASKETBALL	0.50			
11	NUMBER OF HOURS TO MAKE A FOOTBALL	0.30			
12	COST OF LABOR -- 1 MACHINE HOUR	10.00			
13	COST OF MATERIALS -- 1 BASKETBALL	2.00			
14	COST OF MATERIALS -- 1 FOOTBALL	1.25			

FIGURE D-2   CONSTANTS section

- The SELLING PRICE for a single basketball and a single football is shown.
- The TAX RATE is the rate applied to income before taxes to compute income tax expense.
- The NUMBER OF MACHINE HOURS needed to make a basketball and a football is shown. Note that a ball-making machine can produce two basketballs in an hour.
- COST OF LABOR: A ball is made by a worker using a ball-making machine. A worker is paid $10 for each hour he or she works at a machine.
- COST OF MATERIALS: The costs of raw materials for a basketball and football are shown.

Notice that the profit margins (selling price, less costs of labor and materials) for the two products are not the same. They have different selling prices and different inputs (raw materials and hours to make), and the inputs have different costs per unit. Also note that you cannot tell from the data how many hours of the shared resource (machine hours) will be devoted to basketballs and how many will be devoted to footballs, because you don't know in advance how many basketballs and footballs will be made.

CALCULATIONS Section

In the **CALCULATIONS** section, you will calculate intermediate results that (1) will be used in the spreadsheet body and/or (2) will be used as constraints. Before entering formulas, format the Calculations range cells for two decimal places (cell formatting directions were given in Tutorial C). Figure D-3 shows the skeleton and formulas you should enter. A discussion of the cell formulas follows the figure.

N O T E

Cell widths are changed here merely to show the formulas—you need not change the width.

	A	B	C	D
16	**CALCULATIONS**			
17	RATIO OF BASKETBALLS TO FOOTBALLS	=B3/B4		
18	TOTAL BASKETBALL HOURS USED	=B3*B10		
19	TOTAL FOOTBALL HOURS USED	=B4*B11		
20	TOTAL MACHINE HOURS USED (BB + FB)	=B18+B19		

FIGURE D-3   CALCULATIONS section cell formulas

- RATIO OF BASKETBALLS TO FOOTBALLS: This number (cell B17) will be needed in a constraint.
- TOTAL BASKETBALL HOURS USED: The number of machine hours needed to make all basketballs (B3 * B10) is computed in cell B18. Cell B10 has the constant for the hours needed to make one basketball. Cell B3 (a changing cell) has the number of basketballs made. (Currently, this cell shows one ball, but that number will change when the Solver works on the problem.)
- TOTAL FOOTBALL HOURS USED: The number of machine hours needed to make all footballs is calculated similarly in cell B19.
- TOTAL MACHINE HOURS USED (BB + FB): The number of hours needed to make both kinds of balls (cell B20) will be used in constraints; this value is the sum of the hours just calculated for footballs and basketballs.

Notice that constants in the Excel cell formulas in Figure D-3 are referred to by their cell addresses. Use the cell address of a constant rather than hard-coding a number in the Excel expression. If the number must be changed later, you will have to change it only in the **CONSTANTS** section cell, not in every cell formula in which you used the value.

Notice that you do not calculate the amounts in the changing cells (here, the number of basketballs and footballs to produce). The Solver will compute those numbers. Also, notice that you can use the changing cell addresses in your formulas. When you do that, you assume the Solver has put the optimal values in each changing cell; your expression makes use of those numbers.

Figure D-4 shows the calculated values after Excel evaluates the cell formulas (with 1s in the changing cells).

	A	B	C	D	E
16	CALCULATIONS				
17	RATIO OF BASKETBALLS TO FOOTBALLS	1.00			
18	TOTAL BASKETBALL HOURS USED	0.50			
19	TOTAL FOOTBALL HOURS USED	0.30			
20	TOTAL MACHINE HOURS USED (BB + FB)	0.80			

FIGURE D-4 CALCULATIONS section cell values

INCOME STATEMENT Section

The target value is calculated in the spreadsheet body in the **INCOME STATEMENT** section. This is the value that the Solver is expected to maximize or minimize. The spreadsheet body can take any form. In this textbook's Solver cases, the spreadsheet body will be an income statement. Figure D-5 shows the skeleton and formulas that you should enter. A discussion of the line item cell formulas follows the figure.

NOTE

Income statement cells were formatted for two decimal places.

	A	B	C
22	INCOME STATEMENT		
23	BASKETBALL REVENUE (SALES)	=B3*B7	
24	FOOTBALL REVENUE (SALES)	=B4*B8	
25	TOTAL REVENUE	=B23+B24	
26	BASKETBALL MATERIALS COST	=B3*B13	
27	FOOTBALL MATERIALS COST	=B4*B14	
28	COST OF MACHINE LABOR	=B20*B12	
29	TOTAL COST OF GOODS SOLD	=SUM(B26:B28)	
30	INCOME BEFORE TAXES	=B25-B29	
31	INCOME TAX EXPENSE	=IF(B30<=0,0,B30*B9)	
32	NET INCOME AFTER TAXES	=B30-B31	

FIGURE D-5 INCOME STATEMENT section cell formulas

- REVENUE (cells B23 and B24) equals the number of balls times the respective unit selling price. The number of balls is in the changing cells, and the selling prices are constants.
- TOTAL REVENUE is the sum of basketball and football revenue.
- MATERIALS COST (cells B26 and B27) follows a similar logic: number of units times unit cost.
- COST OF MACHINE LABOR is the calculated number of machine hours times the hourly labor rate for machine workers.
- TOTAL COST OF GOODS SOLD is the sum of the cost of materials and the cost of labor.

This is the logic of income tax expense: If INCOME BEFORE TAXES is less than or equal to zero, the tax is zero; otherwise, the income tax expense equals the tax rate times income before taxes. An =IF() statement is needed in cell B31.

Excel evaluates the formulas. Figure D-6 shows the results (assuming 1s in the changing cells).

	A	B	C	D	E
22	**INCOME STATEMENT**				
23	BASKETBALL REVENUE (SALES)	14.00			
24	FOOTBALL REVENUE (SALES)	11.00			
25	TOTAL REVENUE	25.00			
26	BASKETBALL MATERIALS COST	2.00			
27	FOOTBALL MATERIALS COST	1.25			
28	COST OF MACHINE LABOR	8.00			
29	TOTAL COST OF GOODS SOLD	11.25			
30	INCOME BEFORE TAXES	13.75			
31	INCOME TAX EXPENSE	3.85			
32	NET INCOME AFTER TAXES	9.90			

FIGURE D-6   INCOME STATEMENT section cell values

Constraints

Constraints are rules that the Solver must observe when computing the optimal answer to a problem. Constraints need to refer to calculated values, or to values in the spreadsheet body. Therefore, you must build those calculations into the spreadsheet design so they are available to your constraint expressions. (There is no section on the face of the spreadsheet for constraints. You'll use a separate window to enter constraints.)

Figure D-7 shows the English and Excel expressions for the basketball and football production problem constraints. A discussion of the constraints follows the figure.

Expression in English	Excel Expression
TOTAL MACHINE HOURS >= 39000	B20 >= 39000
TOTAL MACHINE HOURS <= 40000	B20 <= 40000
MIN BASKETBALLS = 30000	B3 >= 30000
MAX BASKETBALLS = 60000	B3 <= 60000
MIN FOOTBALLS = 20000	B4 >= 20000
MAX FOOTBALLS = 40000	B4 <= 40000
RATIO BBs TO FBs-MIN = 1.5	B17 >= 1.5
RATIO BBs TO FBs-MAX = 1.7	B17 <= 1.7
NET INCOME MUST BE POSITIVE	B32 >= 0

FIGURE D-7   Solver Constraint expressions

- As shown in Figure D-7, notice that a cell address in a constraint expression can be a cell address in the **CHANGING CELLS** section, a cell address in the **CONSTANTS** section, a cell address in the **CALCULATIONS** section, or a cell address in the spreadsheet body.
- You'll often need to set minimum and maximum boundaries for variables. For example, the number of basketballs (MIN and MAX) varies between 30,000 and 60,000.
- Often a boundary value is zero because you want the Solver to find a non-negative result. For example, here you want only answers that yield a positive net income. You tell the Solver that the amount in the net income cell must equal or exceed zero so the Solver does not find an answer that produces a loss.
- Machine hours must be shared between the two kinds of balls. The constraints for the shared resource are B20 >= 39000 and B20 <= 40000, where cell B20 shows the total hours used to make both basketballs and footballs. The shared resource constraint seems to be the most difficult kind of constraint for students to master when learning the Solver.
- Marketing wants some product balance between products, so ratios of basketballs to footballs must be in the constraints.

Running the Solver: Mechanics

To set up the Solver, you must tell the Solver:

- The cell address of the "target" variable that you are trying to maximize (or minimize, as the case may be).
- The changing cell addresses.
- The expressions for the constraints.

The Solver will record its answers in the changing cells and on a separate sheet.

Beginning to Set Up the Solver

█ AT THE KEYBOARD

To start setting up the Solver, first select the Data tab. In the Analysis group, select the Solver. The first thing you see is a Solver Parameters window, as shown in Figure D-8. Use the Solver Parameters window to specify the target cell, the changing cells, and the constraints. (If you do not see the Solver tool in the Analysis group, you should try to install it. Click the Office Button, then click **Excel Options**, click **Add-Ins**, click the **Go** button for Manage: Excel Add-ins, select the check box for Solver Add-in, and then click **OK**. When prompted to Install, click **Yes**.)

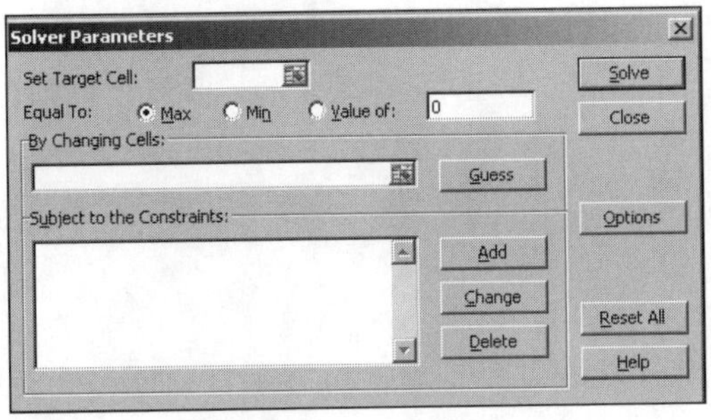

FIGURE D-8   Solver Parameters window

Setting the Target Cell

To set a target cell, use the following procedure:

1. Click the **Set Target Cell** box and enter B32 (net income).
2. Accept Max, the default.
3. Enter a zero for no desired net income value (Value of). Do *not* press Enter when you finish. You'll navigate within this window by clicking in the next input box.

Figure D-9 shows the entering of data in the Set Target Cell box.

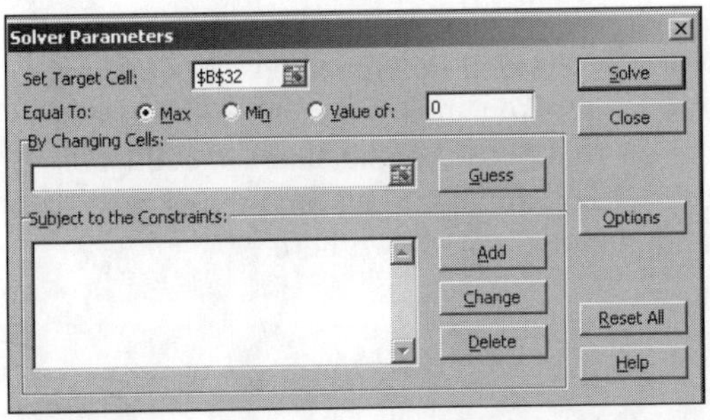

FIGURE D-9   Entering data in the Set Target Cell box

When you enter the cell address, the Solver may put in dollar signs, as if for absolute addressing. Ignore them—do not try to delete them.

Setting the Changing Cells

The changing cells are the cells for the balls, which are in the range of cells B3:B4. Click the **By Changing Cells** box and enter B3:B4, as shown in Figure D-10. (Do *not* press Enter.)

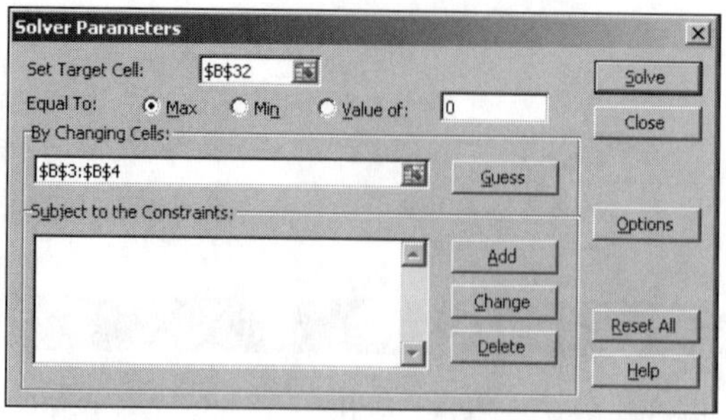

FIGURE D-10   Entering data in the By Changing Cells box

Entering Constraints

You are now ready to enter the constraint formulas one by one. To start, click the **Add** button. As shown in Figure D-11, you'll see the Add Constraint window (shown here with the minimum basketball production constraint entered).

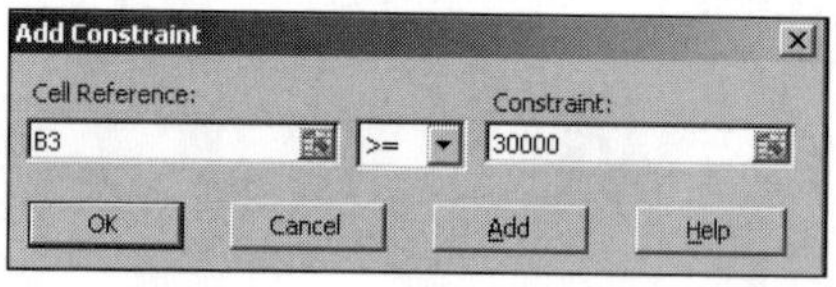

FIGURE D-11   Entering data in the Add Constraint window

You should note the following about entering constraints:

- To enter a constraint expression, do four things: (1) Type the variable's cell address in the left Cell Reference input box; (2) select the operator (<=, =, or >=) in the smaller middle box; (3) enter the expression's right-side value, which is either a raw number or the cell address of a value, into the Constraint box; and (4) click **Add** to enter the constraint into the program. If you change your mind about the expression and do not want to enter it, click **Cancel**.
- The minimum basketballs constraint is B3 >= 30000. Enter that constraint now. (Later, the Solver may put an equal sign in front of the 30000 and dollar signs in the cell reference.)
- After entering the constraint formula, click the **Add** button. Doing so puts the constraint into the Solver model. It also leaves you in the Add Constraint window, allowing you to enter other constraints. You should enter those now. See Figure D-7 for the logic.
- When you're done entering constraints, click the **Cancel** button. That takes you back to the Solver Parameters window.

Referring again to Figure D-11, you should not put an expression in the Cell Reference box. For example, the constraint for the minimum basketball-to-football ratio is B3/B4 >= 1.5. You should not put =B3/B4 in the Cell Reference box. The ratio is computed in the **CALCULATIONS** section of the spreadsheet (in cell B17). When adding that constraint, enter B17 in the Cell Reference box. (Although you are allowed to put an expression in the Constraint box, this technique is not recommended and is not shown here.)

After entering all of the constraints, you'll be back at the Solver Parameters window. You will see that the constraints have been entered into the program. Not all constraints will show, due to the size of the box. The top part of the box's constraints area looks like the portion of the spreadsheet shown in Figure D-12.

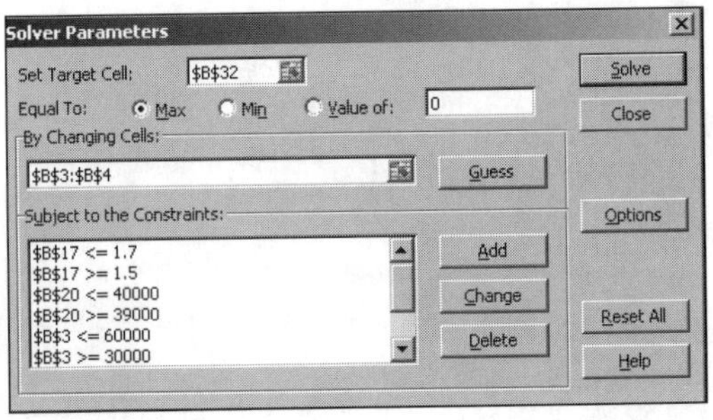

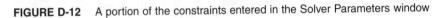

FIGURE D-12   A portion of the constraints entered in the Solver Parameters window

Using the scroll arrow, reveal the rest of the constraints, as shown in Figure D-13.

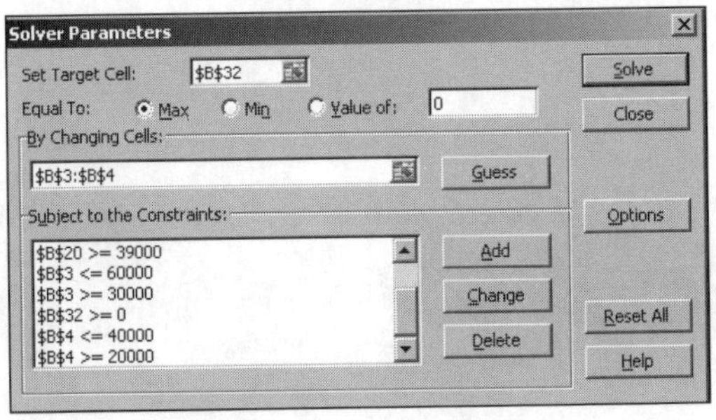

FIGURE D-13   Remainder of constraints entered in the Solver Parameters window

Computing the Solver's Answer

To have the Solver calculate answers, click **Solve** in the upper-right corner of the Solver Parameters window. The Solver does its work in the background—you do not see the internal calculations. Then the Solver gives you a Solver Results window, as shown in Figure D-14.

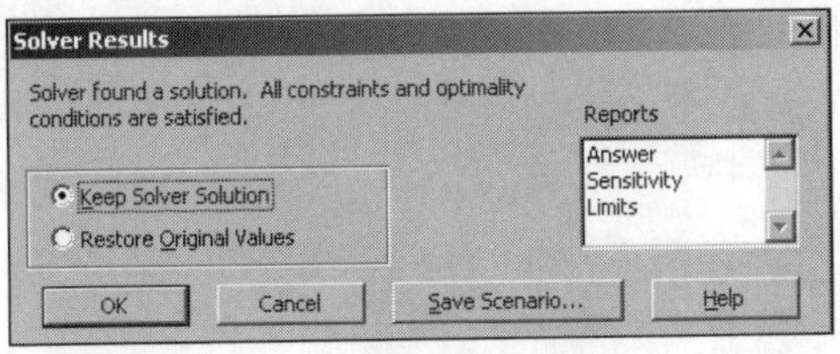

FIGURE D-14   Solver Results window

In the Solver Results window, the Solver tells you that it has found a solution and that the optimality conditions were met. That is a very important message—you should always check for it. It means that an answer was found and the constraints were satisfied.

By contrast, your constraints might be such that the Solver cannot find an answer. For example, suppose you had a constraint that said, in effect, "Net income must be at least a billion dollars." That amount cannot be reached, given so few basketballs and footballs and the prices. The Solver would report that no answer is feasible. The Solver may find an answer by ignoring some constraints. The Solver would tell you that, too. In either case, there would be something wrong with your model and you would need to rework it.

There are two ways to see your answers. One way is to click **OK**, which lets you see the new changing cell values. A more formal (and complete) way is to click **Answer** in the Reports box and then click **OK**. That puts detailed results into a new sheet in your Excel book. The new sheet is called an Answer Report. All answer reports are numbered sequentially as you run the Solver.

To see the Answer Report, click its tab, as shown in Figure D-15 (here, Answer Report 1).

26	B4	NUMBER OF FOOTBALLS	38095.2442 B4>=20000	Not Binding
27	B4	NUMBER OF FOOTBALLS	38095.2442 B4<=40000	Not Binding
28				
29				
30				
31				
32				
33				
34				
35				
36				
37				
38				

Answer Report 1 / Sheet1 / Sheet2 / Sheet3

FIGURE D-15   Answer Report Sheet tab

The top portion of the report is shown in Figure D-16.

	A	B	C	D	E	F
1	Microsoft Excel 12.0 Answer Report					
2	Worksheet: [SPORTS1.xlsx]Sheet1					
3						
4						
5						
6	Target Cell (Max)					
7		Cell	Name	Original Value	Final Value	
8		B32	NET INCOME AFTER TAXES	9.90	473142.87	
9						
10						
11	Adjustable Cells					
12		Cell	Name	Original Value	Final Value	
13		B3	NUMBER OF BASKETBALLS	1	57142.85348	
14		B4	NUMBER OF FOOTBALLS	1	38095.2442	

FIGURE D-16   Top portion of the Answer Report

Figure D-17 shows the remainder of the Answer Report.

	A	B	C	D	E	F
17	Constraints					
18		Cell	Name	Cell Value	Formula	Status
19		B17	RATIO OF BASKETBALLS TO FOOTBALLS	1.50	B17<=1.7	Not Binding
20		B17	RATIO OF BASKETBALLS TO FOOTBALLS	1.50	B17>=1.5	Binding
21		B20	TOTAL MACHINE HOURS USED (BB + FB)	40000.00	B20<=40000	Binding
22		B20	TOTAL MACHINE HOURS USED (BB + FB)	40000.00	B20>=39000	Not Binding
23		B32	NET INCOME AFTER TAXES	473142.87	B32>=0	Not Binding
24		B3	NUMBER OF BASKETBALLS	57142.85348	B3>=30000	Not Binding
25		B3	NUMBER OF BASKETBALLS	57142.85348	B3<=60000	Not Binding
26		B4	NUMBER OF FOOTBALLS	38095.2442	B4>=20000	Not Binding
27		B4	NUMBER OF FOOTBALLS	38095.2442	B4<=40000	Not Binding

FIGURE D-17   Remainder of the Answer Report

At the beginning of this tutorial, the changing cells had a value of 1 and the income was $9.90 (Original Value). The optimal solution values (Final Value) also are shown: $473,142.87 for net income (the target) and 57,142.85 basketballs and 38,095.24 footballs for the changing (adjustable) cells. (Of course, you cannot make a part of a ball. The Solver can be asked to find only integer solutions; that technique is discussed later in this tutorial.)

The report also shows detail for the constraints: the constraint expression and the value that the variable has in the optimal solution. *Binding* means the final answer caused the Solver to bump up against the constraint. For example, the maximum number of machine hours was 40,000, which is the value the Solver used to find the answer.

Not binding means the reverse. A better word for *binding* might be *constraining*. For example, the 60,000 maximum basketball limit did not constrain the Solver.

The procedures used to change (edit) or delete a constraint are discussed later in this tutorial.

Use the Office Button to print the worksheets (Answer Report and Sheet1). Save the Excel file (Save). Then use Save As to make a new file called **SPORTS2.xlsx** to be used in the next section of this tutorial.

EXTENDING THE EXAMPLE

Next, you'll modify the sporting goods spreadsheet. Suppose management wants to know what net income would be if certain constraints were changed. In other words, management wants to play "what if" with certain Base Case constraints. The resulting second case is called the Extension Case. Here are some changes to the original Base Case conditions:

- Assume maximum production constraints will be removed.
- Similarly, the basketball-to-football production ratios (1.5 and 1.7) will be removed.
- There still will be minimum production constraints at some low level. Assume at least 30,000 basketballs and 30,000 footballs will be produced.
- The machine-hours shared resource imposes the same limits as previously.
- A more ambitious profit goal is desired. The ratio of net income after taxes to total revenue should be greater than or equal to .33. This constraint will replace the constraint calling for profits greater than zero.

AT THE KEYBOARD

Begin by putting 1s in the changing cells. You need to compute the ratio of net income after taxes to total revenue. Enter that formula in cell B21. (The formula should have the net income after taxes cell address in the numerator and the total revenue cell address in the denominator.) In the Extension Case, the value of this ratio for the Solver's optimal answer must be at least .33. Click the **Add** button and enter that constraint.

Then in the Solver Parameters window, constraints that are no longer needed are highlighted (click to select) and deleted (click the **Delete** button). Do that for the net income >= 0 constraint, the maximum football and basketball constraints, and the basketball-to-football ratio constraints.

The minimum football constraint must be modified, not deleted. Select that constraint, then click **Change**. That takes you to the Add Constraint window. Edit the constraint so 30,000 is the lower boundary.

When you are finished with the constraints, your Solver Parameters window should look like the one shown in Figure D-18.

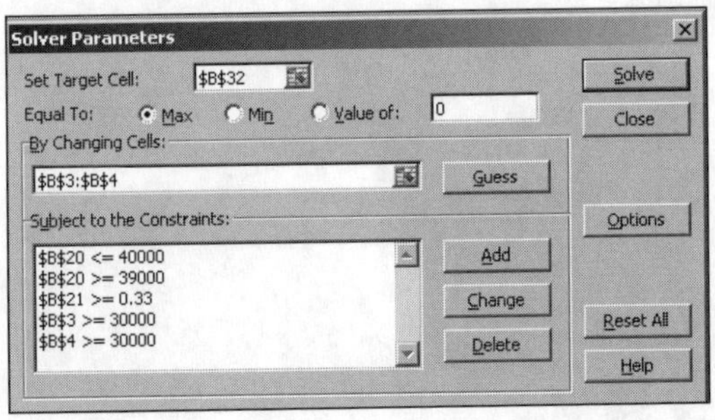

FIGURE D-18 Extension Case Solver Parameters window

You can tell the Solver to solve for integer values. Here, cells B3 and B4 should be whole numbers. You use the Int constraint to do that. Figure D-19 shows how to enter the Int constraint.

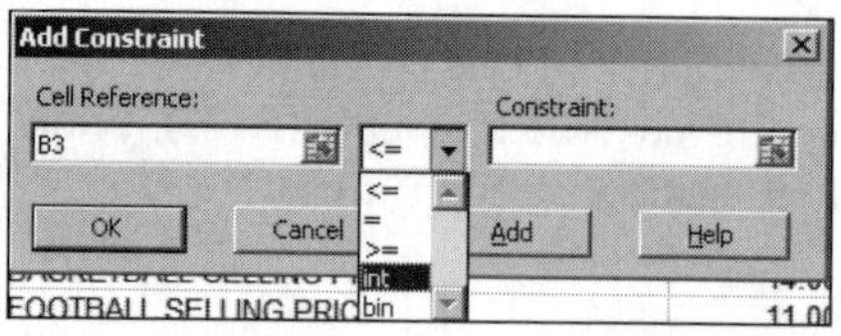

FIGURE D-19   Entering the Int constraint

Make those constraints for the changing cells. Your constraints should now look like the beginning portion of those shown in Figure D-20.

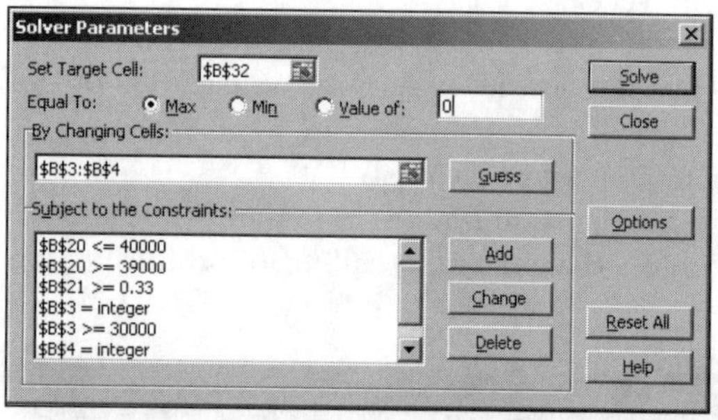

FIGURE D-20   Portion of Extension Case constraints

Scroll to see the remainder of the constraints, as shown in Figure D-21.

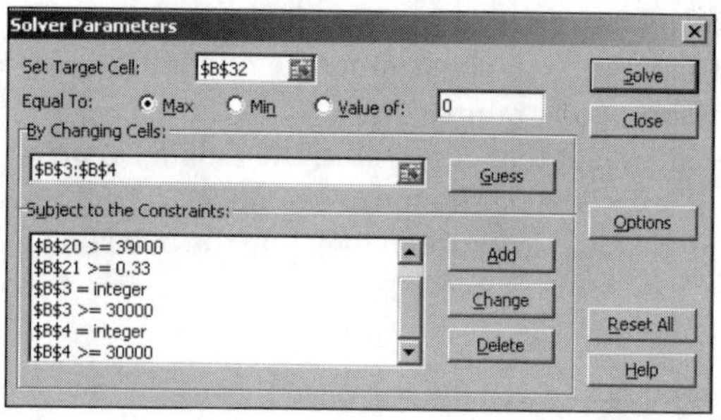

FIGURE D-21   Remainder of Extension Case constraints

The constraints are now only for the minimum production levels, the ratio of net income after taxes to total revenue, machine-hours shared resource constraints, and whole number output. When the Solver is run, the values in the Answer Report look like those shown in Figure D-22.

	A	B	C	D	E	F
Target Cell (Max)						
		Cell	Name	Original Value	Final Value	
		B32	NET INCOME AFTER TAXES	9.90	556198.38	
Adjustable Cells						
		Cell	Name	Original Value	Final Value	
		B3	NUMBER OF BASKETBALLS	1	30000	
		B4	NUMBER OF FOOTBALLS	1	83333	
Constraints						
		Cell	Name	Cell Value	Formula	Status
		B20	TOTAL MACHINE HOURS USED (BB + FB)	39999.90	B20>=39000	Not Binding
		B21	RATIO OF NET INCOME TO TOTAL REVENUE	0.416109655	B21>=0.33	Not Binding
		B20	TOTAL MACHINE HOURS USED (BB + FB)	39999.90	B20<=40000	Not Binding
		B3	NUMBER OF BASKETBALLS	30000	B3>=30000	Binding
		B4	NUMBER OF FOOTBALLS	83333	B4>=30000	Not Binding
		B3	NUMBER OF BASKETBALLS	30000	B3=integer	Binding
		B4	NUMBER OF FOOTBALLS	83333	B4=integer	Binding

FIGURE D-22   Extension Case Answer Report

The Extension Case answer differs from the Base Case answer. Which production schedule should management use: the one that has maximum production limits or the one that has no such limits? That question is posed to get you to think about the purpose of using a DSS program. Two scenarios, the Base Case and the Extension Case, were modeled in the Solver. The very different answers are shown in Figure D-23.

	Base Case	Extension Case
Basketballs	57,143	30,000
Footballs	38,095	83,333

FIGURE D-23   The Solver's answers for the two cases

Can you use this output alone to decide how many of each kind of ball to produce? No, you cannot. You also must refer to the "Target," which in this case is net income. Figure D-24 shows the answers with net income target data.

	Base Case	Extension Case
Basketballs	57,143	30,000
Footballs	38,095	83,333
Net Income	$473,143	$556,198

FIGURE D-24   The Solver's answers for the two cases—with target data

Viewed this way, the Extension Case production schedule looks better because it gives you a higher target net income.

At this point, you should save the SPORTS2.xlsx file and then close it (Office Button—Close).

USING THE SOLVER ON A NEW PROBLEM

Here is a short problem that will let you find out what you have learned about the Excel Solver.

Setting Up the Spreadsheet

Assume you run a shirt manufacturing company. You have two products: (1) polo-style T-shirts and (2) dress shirts with button-down collars. You must decide how many T-shirts and how many button-down shirts to make. Assume you'll sell every shirt you make.

AT THE KEYBOARD

Start a new file called **SHIRTS.xlsx**. Set up a Solver spreadsheet to handle this problem.

CHANGING CELLS Section

Your changing cells should look like those shown in Figure D-25.

	A	B
1	**SHIRT MANUFACTURING EXAMPLE**	
2	**CHANGING CELLS**	
3	NUMBER OF T-SHIRTS	1
4	NUMBER OF BUTTON-DOWN SHIRTS	1

FIGURE D-25 Shirt manufacturing changing cells

CONSTANTS Section

Your spreadsheet should contain the constants shown in Figure D-26. A discussion of constant cells (and some of your company's operations) follows the figure.

	A	B
6	**CONSTANTS**	
7	TAX RATE	0.28
8	SELLING PRICE: T-SHIRT	8.00
9	SELLING PRICE: BUTTON-DOWN SHIRT	36.00
10	VARIABLE COST TO MAKE: T-SHIRT	2.50
11	VARIABLE COST TO MAKE: BUTTON-DOWN SHIRT	14.00
12	COTTON USAGE (LBS): T-SHIRT	1.50
13	COTTON USAGE (LBS): BUTTON-DOWN SHIRT	2.50
14	TOTAL COTTON AVAILABLE (LBS)	13000000
15	BUTTONS PER T-SHIRT	3.00
16	BUTTONS PER BUTTON-DOWN SHIRT	12.00
17	TOTAL BUTTONS AVAILABLE	110000000

FIGURE D-26 Shirt manufacturing constants

- TAX RATE: The rate is .28 on pretax income, but no taxes are paid on losses.
- SELLING PRICE: You sell polo-style T-shirts for $8 and button-down shirts for $36.
- VARIABLE COST TO MAKE: It costs $2.50 to make a T-shirt and $14 to make a button-down shirt. These variable costs are for machine operator labor, cloth, buttons, etc.
- COTTON USAGE: Each polo T-shirt uses 1.5 pounds of cotton fabric. Each button-down shirt uses 2.5 pounds of cotton fabric.
- TOTAL COTTON AVAILABLE: You have only 13 million pounds of cotton on hand to make all of the T-shirts and button-down shirts.
- BUTTONS: Each polo T-shirt has 3 buttons. By contrast, each button-down shirt has 1 button on each collar tip, 8 buttons down the front, and 1 button on each cuff, for a total of 12 buttons. You have 110 million buttons on hand to be used to make all of your shirts.

CALCULATIONS Section

Your spreadsheet should contain the calculations shown in Figure D-27.

	A	B
19	**CALCULATIONS**	
20	RATIO OF NET INCOME TO TOTAL REVENUE	
21	COTTON USED: T-SHIRTS	
22	COTTON USED: BUTTON-DOWN SHIRTS	
23	COTTON USED: TOTAL	
24	BUTTONS USED: T-SHIRTS	
25	BUTTONS USED: BUTTON-DOWN SHIRTS	
26	BUTTONS USED: TOTAL	
27	RATIO OF BUTTON-DOWNS TO T-SHIRTS	

FIGURE D-27   Shirt manufacturing calculations

Calculations (and related business constraints) are discussed next.

- RATIO OF NET INCOME TO TOTAL REVENUE: The minimum return on sales (ratio of net income after taxes divided by total revenue) is .20.
- COTTON USED/BUTTONS USED: You have a limited amount of cotton and buttons. The usage of each resource must be calculated, then used in constraints.
- RATIO OF BUTTON-DOWNS TO T-SHIRTS: You think you must make at least 2 million T-shirts and at least 2 million button-down shirts. You want to be known as a balanced shirtmaker, so you think the ratio of button-downs to T-shirts should be no greater than 4:1. (Thus, if 9 million button-down shirts and 2 million T-shirts were produced, the ratio would be too high.)

INCOME STATEMENT Section

Your spreadsheet should have the income statement skeleton shown in Figure D-28.

	A	B
29	**INCOME STATEMENT**	
30	T-SHIRT REVENUE	
31	BUTTON-DOWN SHIRT REVENUE	
32	TOTAL REVENUE	
33	VARIABLE COSTS: T-SHIRTS	
34	VARIABLE COSTS: BUTTON-DOWNS	
35	TOTAL COSTS	
36	INCOME BEFORE TAXES	
37	INCOME TAX EXPENSE	
38	NET INCOME AFTER TAXES	

FIGURE D-28   Shirt manufacturing income statement line items

The Solver's target is net income, which must be maximized.

Use the table shown in Figure D-29 to write out your constraints before entering them into the Solver.

Expression in English	Fill in the Excel Expression
Net income to revenue	B20 >= 0.20
Ratio of BDs to Ts	B27 <= 4:1 (0.25)
Min T-shirts	_____ >= 2mill
Min button-downs	_____ >= 2mill
Usage of buttons	B26 <= B17
Usage of cotton	B23 <= B14

FIGURE D-29   Logic of shirt manufacturing constraints

When you are finished with the program, print the sheets. Then use the Office Button to Save the file, close it, and choose Exit to leave Excel.

TROUBLESHOOTING THE SOLVER

Use this section to overcome problems with the Solver and to review some Windows file-handling procedures.

Rerunning a Solver Model

Assume you have changed your spreadsheet in some way and want to rerun the Solver to get a new set of answers. (For example, you may have changed a constraint or a formula in your spreadsheet.) Before you click Solve to rerun the Solver, put the number 1 in the changing cells.

Creating Overconstrained Models

It is possible to set up a model that has no logical solution. For example, in the second version of the sporting goods problem, suppose you had specified that at least 1 million basketballs were needed. When you clicked Solve, the Solver would have tried to compute an answer, but then would have admitted defeat by telling you that no feasible solution was possible, as shown in Figure D-30.

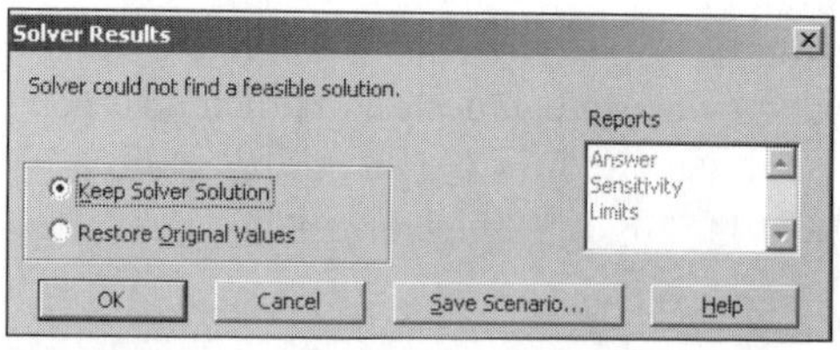

FIGURE D-30   Solver Results message: Solution not feasible

In the Reports window, the choices (Answer, Sensitivity, and Limits) would be in gray, indicating that they are not available as options. Such a model is sometimes referred to as overconstrained.

Setting a Constraint to a Single Amount

You may want an amount to be a specific number, as opposed to a number in a range. For example, if the number of basketballs needed to be exactly 30,000, then the "equals" operator would be selected, as shown in Figure D-31.

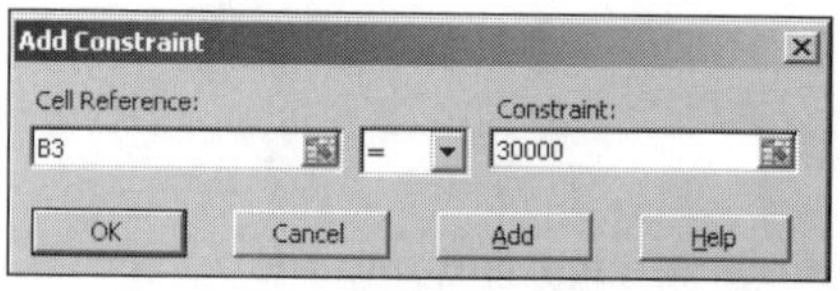

FIGURE D-31   Constraining a value to equal a specific amount

Setting a Changing Cell to an Integer

You may want to force changing cell values to be integers. The way to do that is to select the Int operator in the Add Constraint window. That was described in a prior section.

Forcing the Solver to find only integer solutions slows down the Solver. In some cases, the change in speed can be noticeable to the user. Doing this also can prevent the Solver from seeing a feasible solution when in fact one can be found if the Solver is allowed to find noninteger answers. For those reasons, it's usually best not to impose the integer constraint unless the logic of the problem demands it.

Deleting Extra Answer Sheets

Suppose you've run different scenarios, each time asking for an Answer Report. As a result, you have a number of Answer Report sheets in your Excel file, but you don't want to keep them all. How do you get rid of an Answer Report sheet? Follow this procedure: First, select the Home tab. In the Cells group, click the **Delete** drop-down arrow and select Delete Sheet. You will *not* be asked if you really mean to delete. Therefore, make sure you want to delete the sheet before you act.

Restarting the Solver with All New Constraints

Suppose you wanted to start over with a new set of constraints. In the Solver Parameters window, click **Reset All**. You will be asked if you really mean to do that, as shown in Figure D-32.

FIGURE D-32   Reset options warning query

    If you do, select **OK**. That gives you a clean slate, with all entries deleted, as shown in Figure D-33.

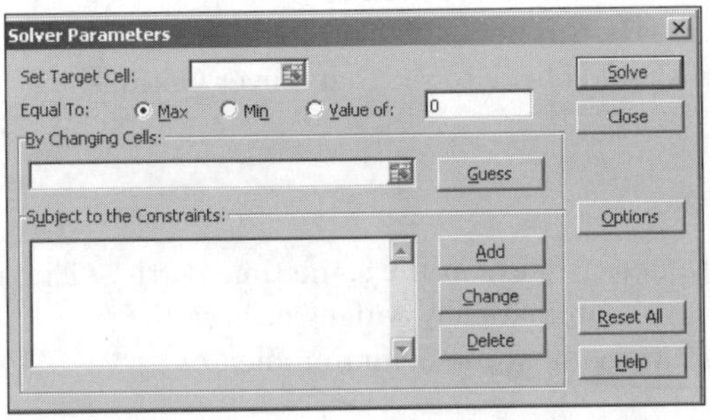

FIGURE D-33   Reset Solver Parameters window

    As you can see, the target cell, changing cells, and constraints have been reset. From this point, you can specify a new model.

NOTE

If you select Reset All, you really are starting over. If you merely want to add, delete, or edit a constraint, do not use Reset All. Use the Add, Delete, or Change buttons as appropriate.

Using the Solver Options Window

The Solver has a number of internal settings that govern its search for an optimal answer. If you click the Options button in the Solver Parameters window, you will see the defaults for those settings, as shown in Figure D-34.

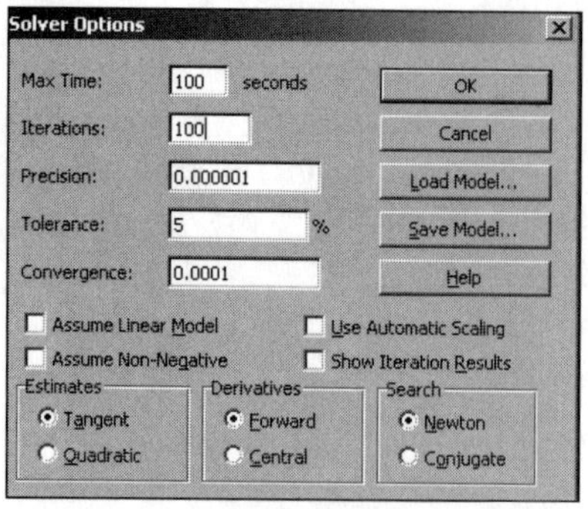

FIGURE D-34   Solver Options window with default settings for Solver Parameters

In general, Solver Options govern how long the Solver works on a problem and/or how precise it must be in satisfying constraints. You should not check Assume Linear Model if changing cells are multiplied or divided (as they are in this book's cases) or if some of the spreadsheet's formulas use exponents.

You should not have to change these default settings for the cases in this book. If you think that your Solver work is correct but the Solver cannot find a feasible solution, you should check to see that Solver Options are set as shown in Figure D-34.

Printing Cell Formulas in Excel

To show the Cell Formulas on the screen, press Ctrl and the left quote (') keys at the same time: Ctrl-'. (The left quote is usually on the same key as the tilde (~).) Pressing Ctrl-' automatically widens cells so the formulas can be read. You can change cell widths by clicking and dragging at the column indicator (A, B, C, etc.) boundaries.

To print the formulas, use the Office Button and select Print. Print the sheet as you would normally. To restore the screen to its typical appearance (showing values, not formulas), press Ctrl-' again. (It's like a toggle switch.) If you did not change any column widths while in the cell formula view, the widths will be as they were.

Reviewing the Printing, Saving, and Exiting Procedures

Print the Solver spreadsheets in the normal way. Activate the sheet, then select the Office Button and Print. You can print an Answer Report sheet the same way.

To save a file, use the Office Button and select Save (or Save As). Make sure you select the proper drive (for example, drive A:) if you intend your file to be on a disk. When exiting from Excel, always start with the Office Button. Then select Close (with the disk in drive A: if your file is on the disk) and select Exit. Only then should you take the disk out of drive A:.

N O T E

If you use File—Exit (without selecting Close first), you risk losing your work.

Sometimes the Solver will come up with strange results. For instance, your results might differ from the target answers that your instructor provides for a case. Thinking that you've done something wrong, you ask to compare your cell formulas and constraint expressions with those your instructor created. Lo and behold, you see no differences! Surprisingly and for no apparent reason, the Solver occasionally produces slightly different outputs from inputs that are seemingly the same. This may occur because for your application, the order of the constraints matters, or even the order in which they are entered matters. In any case, if you are close to the target answers but cannot see any errors, it's best to see your instructor for guidance, rather than to spin your wheels.

Here is another example. Assume you ask for integer changing cell outputs. The Solver may tell you that the correct output is 8.0000001, or 7.9999999. In such situations, the Solver is apparently not sure about its own rounding. In that case, you should merely humor the Solver and (continuing the example) take the result as the integer 8, which is what the Solver was trying to report in the first place.

CASE **8**

THE HOMETOWN UNIVERSITY MENU PLANNING DECISION

Decision Support Using Excel

PREVIEW

Your hometown university wants to design a dormitory meal plan that is tasty, nutritious, and profitable. In this case, you will use the Excel Solver to decide on the best mix of meals to offer students.

PREPARATION

Before attempting this case, you should:

- Review spreadsheet concepts discussed in class and/or in your textbook.
- Complete any exercises that your instructor assigns.
- Complete any part of Tutorial D that your instructor assigns, or refer to it as necessary.
- Review file-saving procedures for Windows programs. These are discussed in Tutorial C.
- Refer to Tutorials E and F as necessary.

BACKGROUND

Your hometown university's dormitory service wants to change the meal plan that it offers to students who live on campus. The goals are to make the meals sufficiently tasty and healthy. Of course, the meal plan must also make money for the hometown university.

Students who sign up for the meal plan would be required to buy 21 meals a week—7 breakfasts, 7 lunches, and 7 dinners. Under the proposed meal plan, the student would buy some "tasty" breakfasts and some "healthy" breakfasts. A "tasty" meal would be somewhat less nutritious than a "healthy" meal, but it would be more pleasing to the taste. The student would also buy some tasty lunches and some healthy lunches, and some tasty dinners and some healthy dinners. A healthy meal would be more nutritious than a tasty meal.

The university wants to decide at the beginning of a semester how many of each kind of meal would be made and served in each week of the semester. The semester meal plan bill for a student would then be computed based on this mix.

Each kind of meal will be priced differently. The university's direct cost of making and serving each kind of meal—in other words, the cost of the recipe items and labor required—would also differ.

Each kind of meal has been rated on the following key nutritional measures:

- Calories—per meal
- Fruit—cups per meal
- Vegetables—cups per meal
- Grain—ounces per meal
- Meat—ounces per meal
- Dairy—ounces per meal
- Fat—teaspoons per meal
- Refined sugar—ounces per meal

Tasty meals differ from healthy meals in several ways. For example, a tasty meal would be higher in fats and/or refined sugar than a healthy meal. A healthy meal would be higher in grains, fruits, and vegetables than a tasty meal. To further explain the difference, a tasty breakfast might include a sweet roll and a few sausage links. A healthy breakfast would eschew the roll and sausage in favor of whole-grain cereal.

Dormitory service cooks have a number of recipes for each kind of meal—for example, there are many different kinds of tasty dinners. Each recipe has been analyzed and the average amount of each nutritional measure has been computed. In other words, the average calories for a tasty breakfast is known, as is the average amount of fruit in a healthy lunch, the average amount of fat per tasty dinner, and so on. Average values for each nutritional measure in each kind of meal are summarized in Figures 8-1 and 8-2. Figure 8-1 shows average calories, fruit, vegetables, and grain per type of meal.

Type of meal	Calories (per meal)	Fruit (cup/meal)	Vegetables (cup/meal)	Grain (ounce/meal)
Healthy breakfast	700	1.00	1.00	3.00
Tasty breakfast	1000	0.50	0.50	1.00
Healthy lunch	800	1.00	1.50	3.00
Tasty lunch	1100	0.50	0.50	1.00
Healthy dinner	1000	1.00	2.00	3.00
Tasty dinner	1250	0.50	0.50	1.00

FIGURE 8-1 Average calories, fruit, vegetables, and grain per type of meal

Figure 8-2 shows average meat, dairy, fat, and refined sugar per type of meal.

Type of meal	Meat (ounce/meal)	Dairy (ounce/meal)	Fat (teaspoon/meal)	Refined sugar (ounce/meal)
Healthy breakfast	1.00	1.00	2.00	0.00
Tasty breakfast	2.00	0.50	4.00	2.00
Healthy lunch	1.50	1.00	2.00	0.00
Tasty lunch	2.50	0.50	4.00	2.00
Healthy dinner	2.50	1.00	2.00	0.00
Tasty dinner	3.50	0.50	4.00	2.00

FIGURE 8-2 Average meat, dairy, fat, and refined sugar per type of meal

The cost of student meal plans will depend on the types of meals chosen. Figure 8-3 shows how much a student will be charged on average for each kind of meal, and how much each meal costs the university to make and serve on average.

Type of meal	Price	Direct cost
Healthy breakfast	3.00	2.00
Tasty breakfast	5.00	3.00
Healthy lunch	5.00	4.00
Tasty lunch	8.00	5.00
Healthy dinner	7.00	6.00
Tasty dinner	10.00	7.00

FIGURE 8-3 Average prices and direct costs of each type of meal

The Hometown University Menu Planning Decision

For the sake of variety, the 21-meal plan must have at least two of each kind of meal each week. A variation of the variety rule is that at least one tasty breakfast must be served per week. Within the plan, exactly 7 breakfasts, 7 lunches, and 7 dinners must be served per week.

University nutritionists have set other rules that are designed to increase overall nutritional value. Total calories in the 21 meals per week must be at least 17,000 but not more than 19,000.

Total fruit in the menu for a week must be at least 13 cups. In other words, the fruit in the 21 meals eaten must add up to at least 13 cups. Total vegetables must be at least 20 cups. Total grains must be at least 49 ounces. Total meat must be at least 37 ounces. Total dairy must be at least 17 cups. Total fats cannot be more than 65 teaspoons. Refined sugar must be less than 14 ounces.

The university's chief financial officer (CFO) has also established a rule that the total price for the 21 meals in a week must be at least $100. In addition, the CFO says that fixed administrative costs over and above making and serving the meals are $20 per week.

The CFO notes that all university operating divisions (such as the dormitory operation) are "taxed" internally. The profit that each division makes for the university is taxed at 10% by the CFO. Tax receipts are used to fund construction projects on campus. Divisions that do not make a profit are not taxed, however.

In the plan, the university requires the student to take all 21 meals in a week. Within the dormitory division, this plan is referred to as the "Base Case." The question in the Base Case is: How many of each of the six types of meals should be served in a week to satisfy nutritional goals and to maximize net income after taxes?

Some managers in the dormitory division think that it is a mistake to require today's busy student to take all 21 meals in a week. These managers think that a student should be required to sign up for (and pay for) most of the 21 meals in a week, but not for all. By this way of thinking, a total of 18 meals would be served in the week and nutritional and financial goals would be adjusted accordingly. This way of thinking about the problem is called the "Extension Case" within the dormitory division. Of course, in this case, the university would still want to maximize net income after taxes.

The CFO and dormitory managers have heard that you know how to use the Excel Solver to help with this sort of problem, and they have called you in. You must model two different situations: the Base Case and the alternative Extension Case. After you have finished the Base Case, you will modify its spreadsheet to create the Extension Case. Comparing the results of the two cases will let you develop "what if" scenarios to help make decisions. Dormitory management will use your results to decide how to plan menus in the coming school year.

ASSIGNMENT 1: CREATING A SPREADSHEET FOR DECISION SUPPORT

In this assignment, you will produce a spreadsheet that models the business decision. In Assignment 1A, you will make a Solver spreadsheet to model the Base Case. In Assignment 1B, you will make a Solver spreadsheet to model the Extension Case. In Assignment 2, you will use the spreadsheet models to develop the information needed to recommend the best meal mix, and then you will document your recommendation in a memorandum to university management. In Assignment 3, you will give your recommendation in an oral presentation.

First, you must create the spreadsheet models of the decision. Your spreadsheets should have the following sections:

- CHANGING CELLS
- CONSTANTS
- CALCULATIONS
- INCOME STATEMENT

Your Solver spreadsheets will also include the decision constraints.

A Base Case spreadsheet skeleton is available to you, so you do not need to type in the skeleton. To access the spreadsheet skeleton, go to your Data files, select Case 8, and then select **MENUS1.xlsx.**

Assignment 1A: Creating the Spreadsheet—Base Case

You will model the meal planning problem described. Your model, when run, will tell management how many of each kind of meal to make and serve to students in a week.

A discussion of each spreadsheet section follows. For each section, you will learn (1) how the section should be set up, and (2) the logic of the sections' cell formulas.

CHANGING CELLS Section

Your spreadsheet should have the changing cells shown in Figure 8-4.

	A	B	C
1	**HOMETOWN UNIVERSITY MENU PLANNING**		
2			BASE
3	**CHANGING CELLS**		**CASE**
4	NUMBER OF HEALTHY BREAKFASTS		1
5	NUMBER OF TASTY BREAKFASTS		1
6	NUMBER OF HEALTHY LUNCHES		1
7	NUMBER OF TASTY LUNCHES		1
8	NUMBER OF HEALTHY DINNERS		1
9	NUMBER OF TASTY DINNERS		1

FIGURE 8-4 CHANGING CELLS section

In this section, you are asking the Solver model to compute how many of each kind of meal should be made and served each week to one student. Start with a "1" in each cell. The Solver will change each "1" as it computes the answer. Of course, a fraction of a meal cannot be served.

CONSTANTS Section

Your spreadsheet should have the constants shown in Figure 8-5. An explanation of the line items follows the figure.

	A	B	C	D	E	F	G
11	**CONSTANTS**		CALORIES	FRUIT	VEGETABLE	GRAIN	MEAT
12	TYPE OF MEAL:		PER MEAL	(CUP)	(CUP)	(OUNCE)	(OUNCE)
13	HEALTHY BREAKFAST		700	1.00	1.00	3.00	1.00
14	TASTY BREAKFAST		1000	0.50	0.50	1.00	2.00
15	HEALTHY LUNCH		800	1.00	1.50	3.00	1.50
16	TASTY LUNCH		1100	0.50	0.50	1.00	2.50
17	HEALTHY DINNER		1000	1.00	2.00	4.00	2.50
18	TASTY DINNER		1250	0.50	0.50	1.00	3.50
19					REFINED		DIRECT
20			DAIRY	FAT	SUGAR	PRICE OF	COST OF
21			(CUP)	(TSPN)	(OUNCE)	A MEAL	A MEAL
22	HEALTHY BREAKFAST		1.00	2.00	0.00	3	2
23	TASTY BREAKFAST		0.50	4.00	2.00	5	3
24	HEALTHY LUNCH		1.00	2.00	0.00	5	4
25	TASTY LUNCH		0.50	4.00	2.00	8	5
26	HEALTHY DINNER		1.00	2.00	0.00	7	6
27	TASTY DINNER		0.50	4.00	2.00	10	7
28	MIN DINING PLAN CHARGE		100				
29	FIXED DINING COSTS FOR A WEEK		20				
30	INTERNAL TAX RATE		0.10				

FIGURE 8-5 CONSTANTS section

- CALORIES PER MEAL: Average calories for each type of meal are shown.
- FRUIT (CUP): Average cups of fruit for each type of meal are shown.
- VEGETABLE (CUP): Average cups of vegetables for each type of meal are shown.
- GRAIN (OUNCE): Average ounces of grain for each type of meal are shown.
- MEAT (OUNCE): Average ounces of meat for each type of meal are shown.
- DAIRY (CUP): Average cups of dairy for each type of meal are shown.
- FAT (TSPN): Average teaspoons of fat for each type of meal are shown.
- REFINED SUGAR (OUNCE): Average ounces of refined sugar for each type of meal are shown.
- PRICE OF A MEAL: The price charged to the student for each type of meal is shown.
- DIRECT COST OF A MEAL: The direct cost to make and serve each type of meal is shown.

- MIN DINING PLAN CHARGE: The student must be charged at least $100 a week.
- FIXED DINING COSTS PER WEEK: Fixed administrative costs are $20 a week per student.
- INTERNAL TAX RATE: The university taxes profitable operations at a 10% rate to fund construction projects on campus. If an operation has losses, however, the tax is zero.

CALCULATIONS Section

Your spreadsheet should calculate the amounts shown in Figure 8-6. They will be used in the INCOME STATEMENT section and/or the constraints. Calculated values are based either on changing cell values, constants, and/or other calculated values. An explanation of the line items follows the figure.

	A	B	C
32	**CALCULATIONS**		
33	REVENUE FROM MEALS:		--
34	HEALTHY BREAKFAST		
35	TASTY BREAKFAST		
36	HEALTHY LUNCH		
37	TASTY LUNCH		
38	HEALTHY DINNER		
39	TASTY DINNER		
40	DIRECT COSTS OF MEALS:		--
41	HEALTHY BREAKFAST		
42	TASTY BREAKFAST		
43	HEALTHY LUNCH		
44	TASTY LUNCH		
45	HEALTHY DINNER		
46	TASTY DINNER		
47	NUMBER OF BREAKFASTS SERVED		
48	NUMBER OF LUNCHES SERVED		
49	NUMBER OF DINNERS SERVED		
50	NUMBER OF MEALS SERVED		
51	TOTAL CALORIES		
52	TOTAL FRUIT		
53	TOTAL VEGETABLE		
54	TOTAL GRAIN		
55	TOTAL MEAT		
56	TOTAL DAIRY		
57	TOTAL FAT		
58	TOTAL SUGAR		

FIGURE 8-6 CALCULATIONS section

- REVENUE FROM MEALS: The revenue for each type of meal is a function of the number of that meal served in the week and the price of that type of meal.
- DIRECT COSTS OF MEALS: The direct cost of making and serving each type of meal is a function of the number of that meal served in the week and the direct cost of that type of meal.
- NUMBER OF BREAKFASTS SERVED: This is the total of tasty breakfasts and healthy breakfasts served in the week.
- NUMBER OF LUNCHES SERVED: This is the total of tasty lunches and healthy lunches served in the week.
- NUMBER OF DINNERS SERVED: This is the total of tasty dinners and healthy dinners served in the week.
- NUMBER OF MEALS SERVED: This is the total of all meals served in the week.
- TOTAL CALORIES: This is a function of the number of each type of meal served in the week and the average calories for that type of meal.
- TOTAL FRUIT: This is a function of the number of each type of meal served in the week and the average cups of fruit for that type of meal.
- TOTAL VEGETABLE: This is a function of the number of each type of meal served in the week and the average cups of vegetables for that type of meal.

- TOTAL GRAIN: This is a function of the number of each type of meal served in the week and the average ounces of grain for that type of meal.
- TOTAL MEAT: This is a function of the number of each type of meal served in the week and the average ounces of meat for that type of meal.
- TOTAL DAIRY: This is a function of the number of each type of meal served in the week and the average cups of dairy for that type of meal.
- TOTAL FAT: This is a function of the number of each type of meal served in the week and the average teaspoons of fat for that type of meal.
- TOTAL SUGAR: This is a function of the number of each type of meal served in the week and the average ounces of refined sugar for that type of meal.

INCOME STATEMENT Section

Compute the dining operation's net income after taxes, as shown in Figure 8-7. This statement shows net income for one student's meals in a week. An explanation of the line items follows the figure.

	A	B	C
60	**INCOME STATEMENT**		
61	MEAL PLAN REVENUE		
62	DIRECT COSTS		
63	FIXED DINING COSTS		
64	INCOME BEFORE INCOME TAXES		
65	INCOME TAX EXPENSE		
66	NET INCOME AFTER TAXES		

FIGURE 8-7 INCOME STATEMENT section

- MEAL PLAN REVENUE: This is the sum of revenues calculated for each kind of meal.
- DIRECT COSTS: This is the sum of direct costs calculated for each kind of meal.
- FIXED DINING COSTS: This value is a constant, which can be echoed to here.
- INCOME BEFORE INCOME TAXES: This is the difference between meal plan revenue and the sum of direct costs and fixed dining costs.
- INCOME TAX EXPENSE: Taxes are zero if Income Before Income Taxes is zero or negative. Otherwise, apply the internal tax rate (a constant) to the Income Before Income Taxes to compute the income tax expense.
- NET INCOME AFTER TAXES: This is the difference between Income Before Income Taxes and Income Tax Expense.

Constraints and Running the Solver

In this part of the assignment, you determine the decision constraints for the spreadsheet. Enter the Base Case decision constraints using the Solver. Run the Solver; when it says a solution has been found that satisfies the constraints, ask for the Answer Report.

When you finish, print the entire workbook, including the Answer Report sheet. Save the workbook for the last time (Office Button|Save). Then, to prepare for the Extension Case, use Office Button|Save As to make a new spreadsheet file (**MENUS2.xlsx** would be a good name).

Assignment 1B: Creating the Spreadsheet—Extension Case

University management wants you to model the Extension Case in the Solver. In this case, 18 meals would be provided to students, not 21.

Breakfast is considered the most important meal by dormitory division nutritionists, and so 7 breakfasts must be in the plan. Five or more lunches and 5 or more dinners would be provided. Seven breakfasts, 7 lunches, or 7 dinners could be provided, but obviously not 7 of each, as only 18 meals would be in the plan.

At least one of each of the six kinds of meals would be included in the plan. The upper limit on each of the six kinds of meals would be six.

Nutritional goals would be adjusted for the change in plan. Total calories in the meals per week must be at least 12,070 but not more than 16,300.

Total fruit in the menu for a week must be at least 9 cups. Total vegetables must be at least 14 cups. Total grains must be at least 35 ounces. Total meat must be at least 26 ounces. Total dairy must be at least 12 cups. Total fats cannot be more than 56 teaspoons. There must be some refined sugar, but not more than 12 ounces.

The university's CFO says the total cost of the 18 meals in a week must be at least $80. Fixed administrative costs per week per student would remain at $20.

Modify the Extension Case spreadsheet and related constraints to reflect this way of thinking about the meal plan. Run the Solver. Ask for an Answer Report when the Solver says a solution has been found that satisfies the constraints. When you finish, print the entire workbook, including the Solver Answer Report. Save the file when you finish, close the file, and exit from Excel.

ASSIGNMENT 2: USING THE SPREADSHEET FOR DECISION SUPPORT

You have built the Base Case and Extension Case models because you want to know the profitability of each meal plan. You will now complete your work by: (1) using the worksheets and Answer Reports to gather data, and (2) documenting your analysis and recommendation in a memorandum.

Assignment 2A: Using the Spreadsheets to Gather Data

You have printed the spreadsheet and Answer Report for each case so you can see the results of each approach. Management wants to know how net income after taxes differs between the models. They also want to know the number of each kind of meal in the two plans.

Dormitory operation managers say they would not mind reducing the number of weekly meals from 21 to 18. That would give them more time for sanitation and other housekeeping tasks. The CFO says that if the Extension Case net income after taxes is 90% or more of Base Case net income after taxes, dormitory management and the CFO would seriously consider using the Extension Case. Otherwise, management would opt for the Base Case.

You should summarize the key data in a table that will be included in your memorandum. The form of the table is shown in Figure 8-8.

	Base Case	Extension Case
Healthy breakfast		
Tasty breakfast		
Healthy lunch		
Tasty lunch		
Healthy dinner		
Tasty dinner		
Net income after taxes		

FIGURE 8-8 Format of table to insert into memorandum

You can find the number of each kind of meal served in the Answer Reports and in the worksheet changing cells. You can find the net income after taxes in the Answer Reports and in the worksheets.

Assignment 2B: Documenting Your Recommendation in a Memorandum

Use MS Word to write a brief memorandum to the CFO and dormitory division management about the results of your analysis. Observe the following requirements:

- Your memorandum should be set up as discussed in Tutorial E.
- You need not provide background—the CFO and management are aware of the situation. You should briefly state your analytical method and state the results. Give the CFO and management your recommendation.
- Support the recommendation graphically by including a summary table in your Word memo, as shown in Figure 8-8. Insert the table by following the procedure described in Tutorial E.

ASSIGNMENT 3: GIVING AN ORAL PRESENTATION

Your instructor may request that you also present your analysis and recommendation in an oral presentation. If so, assume that the CFO and dormitory division management have accepted your analysis and recommendation. They have asked you to give a presentation explaining your recommendation to the university's senior officials. Prepare to explain your analysis and recommendation to the group in 10 minutes or less. Use visual aids or handouts that you think are appropriate. Tutorial F has guidance on how to prepare and give an oral presentation.

DELIVERABLES

Assemble the following deliverables for your instructor:

- A printout of your memorandum
- Spreadsheet printouts
- A disk that has your Word memorandum and Excel spreadsheet files. (Do not provide a CD, which would be read-only.)

Staple the printouts together, with the memorandum on top.

THE HOMETOWN CHRISTMAS TREE PRODUCTION DECISION

Decision Support Using Excel

PREVIEW

You are starting a Christmas tree farm and need to decide how many acres to devote to spruce, pine, and Douglas fir trees to maximize profits. In this case, you will use the Excel Solver to decide on the best acreage mix.

PREPARATION

Before attempting this case, you should:

- Review spreadsheet concepts discussed in class and/or in your textbook.
- Complete any exercises that your instructor assigns.
- Complete any part of Tutorial D that your instructor assigns, or refer to it as necessary.
- Review file-saving procedures for Windows programs. These are discussed in Tutorial C.
- Refer to Tutorials E and F as necessary.

BACKGROUND

You have just inherited 250 acres of gently rolling farmland. You and your wife work for a local bank. You both want to retire and devote your time to Christmas tree farming on the land you have inherited. You want to run the farm in the most profitable way possible. You have intensively researched the Christmas tree industry and have learned a great deal.

The United States has many kinds of Christmas trees. In your part of the country, the spruce, pine, and Douglas fir are most popular, and you want to plant those species.

Trees are planted as small shoots. The cost of shoots differs by tree species. Some species need more space in the ground than others. Survival rates differ by species as well.

As trees grow, the land must be cared for—the ground must be loosened occasionally and weeds removed. Herbicides and pesticides must be applied to the ground and trees during growth.

Trees must be groomed periodically during growth to attain an attractively tapered Christmas-tree look. Grooming is called "shearing."

You will be able to maintain the farm with a small permanent work force until harvesting. Then, you will need to hire many more workers. Hours will be long during the harvest, which occurs in the 30 days before Christmas. You think that 35 workers must be hired to work 15 hours a day during the harvest.

Customers would come to the tree farm starting in early November to select and tag their tree and to choose the day they would return to get their cut tree. In the month before Christmas, customers would return to pick up their trees and pay for them.

You need to determine how many acres of spruce, pine, and Douglas firs to plant. You want to maximize net income after income taxes, of course. You think that at least 60 acres of each kind of tree should be planted, and that no more than 100 acres of each type should be planted. Demand for Christmas trees seems to be growing in your area, and you think that all trees harvested will be sold to the public.

Your "base case" is to plant and harvest all 250 acres. However, your wife and daughters have an alternative idea that they call the "extension case." They like working with the arts and crafts that have become so much a part of the American Christmas season—the fancy ornaments, wreathes, and other items with which people decorate their homes during the season. Your wife and daughters would like to build a crafts center. Local craftsmen would drop off their wares on consignment, and the wares would be sold during the harvesting season.

Your wife and daughters would set aside a corner of the center for selling cider, hot chocolate, Christmas cookies, and so on. Five acres, including a parking lot, would be required for the entire operation.

You and your wife have analyzed the economics of running the craft shop for the season. It appears that the shop's operating expenses would exceed sales by $200,000. This pretax loss would effectively be an additional operating expense of the Christmas tree farm. However, you think that the goodwill and publicity generated by the center would attract more people to your tree farm, which would increase demand for all products to some extent. Thus, you think the extra goodwill and publicity would allow you to increase the selling price of your Christmas trees. On that basis, therefore, the craft shop might be worth taking an operating loss. If the extension case's net income after taxes is close to the base case's net income after taxes, you will build and operate the craft center to please your wife and daughters. But if the base case's net income after taxes is clearly superior, you would adopt that strategy.

You know how to use the Excel Solver to help with this sort of problem. You must model two different situations: the Base Case and the alternative Extension Case. After you have finished the Base Case, you will modify its spreadsheet to create the Extension Case. Comparing the results of the two cases will let you develop "what if" scenarios to help make decisions.

ASSIGNMENT 1: CREATING A SPREADSHEET FOR DECISION SUPPORT

In this assignment, you will produce a spreadsheet that models the business decision. In Assignment 1A, you will make a Solver spreadsheet to model the Base Case. In Assignment 1B, you will make a Solver spreadsheet to model the Extension Case. In Assignment 2, you will use the spreadsheet models to develop the information needed to recommend the best tree acreage mix, and then you will document your recommendation in a memorandum. In Assignment 3, you will give your recommendation in an oral presentation to your tax advisor.

First, you must create the spreadsheet models of the decision. Your spreadsheets should have the following sections:

- CHANGING CELLS
- CONSTANTS
- CALCULATIONS
- INCOME STATEMENT

Your Solver spreadsheets will also include the decision constraints.

A Base Case spreadsheet skeleton is available to you, so you do not need to type in the skeleton. To access the spreadsheet skeleton, go to your Data files, select Case 9, and then select **TREES1.xlsx**.

Assignment 1A: Creating the Spreadsheet—Base Case

You will model the acreage mix planning problem described. Your model, when run, will tell you how many acres of each kind of tree to plant.

A discussion of each spreadsheet section follows. For each section, you will learn (1) how the section should be set up, and (2) the logic of the sections' cell formulas.

CHANGING CELLS Section

Your spreadsheet should have the changing cells shown in Figure 9-1.

	A	B	C	D	E
1	HOMETOWN CHRISTMAS TREE PRODUCTION DECISION				
2			BASE		
3	CHANGING CELLS		CASE		
4	NUMBER OF ACRES -- SPRUCE		1		
5	NUMBER OF ACRES -- PINE		1		
6	NUMBER OF ACRES -- DOUGLAS FIR		1		

FIGURE 9-1 CHANGING CELLS section

In this section, you are asking the Solver model to compute how many acres of each kind of tree should be planted. Start with a "1" in each cell. The Solver will change each "1" as it computes the answer. Assume that a fraction of an acre would not be planted.

CONSTANTS Section

Your spreadsheet should have the constants shown in Figures 9-2 and 9-3. An explanation of the line items follows the figures.

	A	B	C
8	**CONSTANTS**		
9	TREES PER ACRE:		--
10	SPRUCE		1,140
11	PINE		1,080
12	DOUGLAS FIR		1,100
13	COST OF PLANTING PER ACRE:		--
14	SPRUCE		300
15	PINE		250
16	DOUGLAS FIR		275
17	HERBICIDE & PESTICIDE COST PER ACRE:		--
18	SPRUCE		215
19	PINE		300
20	DOUGLAS FIR		225
21	COST OF SHEARING PER TREE:		--
22	SPRUCE		0.40
23	PINE		0.30
24	DOUGLAS FIR		0.35
25	COST OF LAND CARE PER ACRE (ALL):		100
26	REVENUE PER TREE:		--
27	SPRUCE		21
28	PINE		20
29	DOUGLAS FIR		22

FIGURE 9-2 CONSTANTS section

- TREES PER ACRE: An acre of your land will yield 1,140 spruce trees, 1,080 pine trees, and 1,100 Douglas fir trees.
- COST OF PLANTING PER ACRE: The cost of planting includes the cost of seedlings, labor, machine usage, and so forth. Costs per acre differ for the three species, as shown.
- HERBICIDE & PESTICIDE COST PER ACRE: Weed and pest killer costs differ by species, as shown.
- COST OF SHEARING PER TREE: The cost of shearing differs by species type. (Note that the costs shown here are by tree, not by acre.)
- COST OF LAND CARE PER ACRE (ALL): The cost of tending to the land is the same for each species: $100 per acre.
- REVENUE PER TREE: The expected selling price for each kind of tree is shown here.

The remaining constants are shown in Figure 9-3:

	A	B	C
30	FIXED COST PER ACRE		200.00
31	INCOME TAX RATE		0.20
32	NUMBER OF HARVEST DAYS		30
33	NUMBER OF HOURS PER DAY		15
34	NUMBER OF WORKERS AVAILABLE		35
35	MINUTES TO HARVEST A TREE		3
36	LABOR COST PER HOUR		15

FIGURE 9-3 CONSTANTS section, continued

- FIXED COST PER ACRE: You will employ a small staff of workers until tree harvesting. Also, there will be operating and administrative expenses throughout. These costs are expected to be $200 an acre.
- INCOME TAX RATE: You expect to pay 20% of your gross profit to local, state, and federal tax authorities.
- NUMBER OF HARVEST DAYS: The harvesting season lasts 30 days.
- NUMBER OF HOURS PER DAY: During the harvest, workers will be in the field 15 hours a day.
- NUMBER OF WORKERS AVAILABLE: You will hire an extra 35 workers during the harvesting season.
- MINUTES TO HARVEST A TREE: A worker needs three minutes to harvest a tree.
- LABOR COST PER HOUR: The 35 workers hired for the harvest will be paid $15 an hour.

CALCULATIONS Section

Your spreadsheet should calculate the amounts shown in Figure 9-4. They will be used in the INCOME STATEMENT section and/or the constraints. Calculated values are based either on changing cell values, constants, and/or other calculated values. An explanation of the line items follows the figure.

	A	B	C
38	**CALCULATIONS**		
39	TOTAL ACRES PLANTED		
40	FIXED COSTS		
41	LAND CARE COSTS		
42	PLANTING COSTS:		--
43	SPRUCE		
44	PINE		
45	DOUGLAS FIR		
46	NUMBER OF TREES HARVESTED:		--
47	SPRUCE		
48	PINE		
49	DOUGLAS FIR		
50	REVENUE:		--
51	SPRUCE		
52	PINE		
53	DOUGLAS FIR		
54	COST OF HERBICIDE & PESTICIDE:		--
55	SPRUCE		
56	PINE		
57	DOUGLAS FIR		
58	COST OF SHEARING:		--
59	SPRUCE		
60	PINE		
61	DOUGLAS FIR		
62	HARVESTING MINUTES AVAILABLE		
63	MAX TREES THAT CAN BE HARVESTED		
64	TOTAL NUMBER OF TREES HARVESTED		
65	HARVESTING HOURS WORKED		
66	TOTAL LABOR COST		

FIGURE 9-4 CALCULATIONS section

- TOTAL ACRES PLANTED: This is the total acreage planted for all species. The changing cells show how many acres of each are planted. The total acres planted cannot exceed 250.
- FIXED COSTS: This is a function of the fixed cost per acre and the total acres planted.
- LAND CARE COSTS: This is a function of the land care cost per acre and the total acres planted.
- PLANTING COSTS: This is a function of the planting cost per acre for the species and the number of acres planted for the species.
- NUMBER OF TREES HARVESTED: This is a function of the number of acres planted for a species and the trees per acre yielded for that species.

- REVENUE: This is a function of the number of trees harvested for a species and revenue per tree (selling price) for the species.
- COST OF HERBICIDE & PESTICIDE: This is a function of the number of acres planted for a species and the herbicide and pesticide cost per acre for that species.
- COST OF SHEARING: This is a function of the number of trees harvested for a species and the shearing cost per tree for that species.
- HARVESTING MINUTES AVAILABLE: This is a function of the number of harvest days, number of harvest workers available, and the number of hours worked per day. Of course, there are 60 minutes in an hour.
- MAX TREES THAT CAN BE HARVESTED: This is a function of the number of harvesting minutes available and the number of minutes that are needed to harvest a tree.
- TOTAL NUMBER OF TREES HARVESTED: This is the sum of calculated values for trees harvested for each species. The number of trees harvested cannot exceed the maximum amount shown in the preceding cell.
- HARVESTING HOURS WORKED: This is a function of the total number of trees harvested, the minutes needed to harvest a tree, and minutes per hour (60).
- TOTAL LABOR COST: This is a function of the harvesting hours worked and labor cost per hour.

INCOME STATEMENT Section

Compute the Christmas tree operation's net income after taxes, as shown in Figure 9-5. This statement shows net income for all trees harvested and sold in the season. An explanation of the line items follows the figure.

	A	B	C
69	**INCOME STATEMENT**		
70	REVENUE		
71	COSTS:		
72	PLANTING		
73	HERBICIDE & PESTICIDE		
74	SHEARING		
75	FIXED		
76	LAND CARE		
77	LABOR		
78	TOTAL COSTS		
79	INCOME BEFORE INCOME TAXES		
80	INCOME TAX EXPENSE		
81	NET INCOME AFTER INCOME TAXES		

FIGURE 9-5 INCOME STATEMENT section

- REVENUE: This is the sum of revenues for each species.
- PLANTING COST: This is the sum of planting costs for each species.
- HERBICIDE & PESTICIDE COST: This is the sum of herbicide and pesticide costs for each species.
- SHEARING COST: This is the sum of shearing costs for each species.
- FIXED COST: This amount was calculated elsewhere and can be echoed here.
- LAND CARE COST: This amount was calculated elsewhere and can be echoed here.
- LABOR COST: This amount was calculated elsewhere and can be echoed here.
- TOTAL COSTS: This is the sum of planting, herbicide & pesticide, shearing, fixed, land care, and labor costs.
- INCOME BEFORE INCOME TAXES: This is the difference between revenue and total costs.
- INCOME TAX EXPENSE: Taxes are zero if Income Before Income Taxes is zero or negative. Otherwise, apply the income tax rate, a constant, to the Income Before Income Taxes to compute the income tax expense.
- NET INCOME AFTER INCOME TAXES: This is the difference between Income Before Income Taxes and Income Tax Expense.

Constraints and Running the Solver

In this part of the assignment, you determine the decision constraints for the spreadsheet. Enter the Base Case decision constraints using the Solver. Run the Solver; when it says a solution has been found that satisfies the constraints, ask for the Answer Report.

When you finish, print the entire workbook, including the Answer Report sheet. Save the workbook for the last time (Office Button|Save). Then, to prepare for the Extension Case, use Office Button|Save As to make a new spreadsheet file (**TREES2.xlsx** would be a good name).

Assignment 1B: Creating the Spreadsheet—Extension Case

You want to model the Extension Case in the Solver. In this case, 5 of the 250 acres would be set aside for the craft center. The income statement should be changed to include an additional operating cost for the $200,000 pretax loss from running the center. Christmas tree prices would change to $22 for spruce trees, $21 for pines, and $23 for Douglas firs.

You would adopt the Extension Case if its net income after taxes was superior to that of the Base Case. You want to please your wife and daughters, and you think business diversification is a good thing, so you would adopt the Extension Case as your strategy if its net income after taxes was within $50,000 of base case net income after taxes.

Modify the Extension Case spreadsheet and related constraints to reflect this way of thinking about the business. Run the Solver. Ask for an Answer Report when the Solver says a solution has been found that satisfies the constraints. When you finish, print the entire workbook, including the Solver Answer Report. Save the file when you finish, close the file, and exit from Excel.

ASSIGNMENT 2: USING THE SPREADSHEET FOR DECISION SUPPORT

You have built the Base Case and Extension Case models because you want to know the profitability of each tree farm strategy. You will now complete your work by: (1) using the worksheets and Answer Reports to gather data, and (2) documenting your analysis and recommendation in a memorandum.

Assignment 2A: Using the Spreadsheets to Gather Data

You have printed the spreadsheet and Answer Report for each case so you can see the results of each approach. You want to know how net income after taxes differs between the two cases. You also want to know the number of acres planted for each kind of tree in each case.

You should summarize the key data in a table that will be included in your memorandum. The form of the table is shown in Figure 9-6.

	Base Case	Extension Case
Number of spruce acres planted		
Number of pine acres planted		
Number of Douglas fir acres planted		
Net income after taxes		

FIGURE 9-6   Format of table to insert into memorandum

You can find the number of acres for each species of tree in the Answer Reports and in the worksheet changing cells. You can find the net income after taxes in the Answer Reports and in the worksheets.

Assignment 2B: Documenting Your Recommendation in a Memorandum

Use MS Word to write a brief memorandum to your income tax preparer, who is also your personal financial advisor, about the results of your analysis. Observe the following requirements:

- Your memorandum should be set up as discussed in Tutorial E.
- You need not provide background—your tax advisor is aware of the situation. You should briefly state your analytical method and state the results. Tell the advisor which strategy you have decided to adopt and why.
- Support the recommendation graphically by including a summary table in your Word memo, as shown in Figure 9-6. Insert the table by following the procedure described in Tutorial E.

ASSIGNMENT 3: GIVING AN ORAL PRESENTATION

Your instructor may request that you also present your analysis and recommendation in an oral presentation. If so, assume that the tax advisor has accepted your analysis and recommendation. The advisor wants the other partners in her accounting firm to learn from the analysis. Prepare to explain your analysis and recommendation to the group in 10 minutes or less. Use visual aids or handouts that you think are appropriate. Tutorial F has guidance on how to prepare and give an oral presentation.

DELIVERABLES

Assemble the following deliverables for your instructor:

- A printout of your memorandum
- Spreadsheet printouts
- A disk that has your Word memorandum and Excel spreadsheet files. (Do not provide a CD, which would be read-only.)

Staple the printouts together, with the memorandum on top.

DECISION SUPPORT CASES
USING BASIC EXCEL FUNCTIONALITY

CASE **10**

THE PORTFOLIO RECONSTRUCTION PROBLEM

Decision Support Using Excel

PREVIEW

You have been asked to spruce up an investor's poorly maintained common stock portfolio. In this case, you will use Excel to identify stocks to keep and stocks to sell.

PREPARATION

- Review spreadsheet concepts discussed in class and/or in your textbook.
- Complete any exercises that your instructor assigns.
- Complete any parts of Tutorials C and D that your instructor assigns, or refer to them as necessary.
- Review file-saving procedures for Windows programs. These are discussed in Tutorial C.
- Refer to Tutorials E and F as necessary.

BACKGROUND

You work for PrimeCommerce Trust Bank. PrimeCommerce's banking business manages money for very wealthy people. Clients turn over huge sums of money to the bank to be invested. The bank's "wealth managers" invest the money, seeking high returns at reasonable levels of risk.

You are one of the bank's younger wealth managers, having worked your way up through various financial analysis positions in the bank. The wealth management position requires a high degree of investment management skill and a high degree of discretion. It's an excellent job and you love it.

A senior wealth manager is retiring. Management is parceling out his accounts to junior members, and you have been given Mrs. J. K. Atkins' account.

Mrs. Atkins is a wealthy heiress. Her husband died young and rich, and she never remarried. She is still active in the community at the age of 75. Mrs. Atkins is very sharp, but never has had much interest in money management. PrimeCommerce has generated enough money for her to run her estate and her life. As a result, she has never questioned the way your predecessor managed her account.

You have reviewed the account status and you are shocked. The account has 200 common stock investments, at least half of which appear to be losing money! A basic tenet of conservative money management at the bank is to keep your winners and cut your losses. In other words, if a stock investment does not work out, sell the stock and buy another one. You cannot believe that your predecessor held on to so many losing stock positions. Your manager, an executive vice president at the bank, is also appalled.

Your goal is to reconstruct the Atkins portfolio. In the reconstruction, you will keep good common stocks. You will sell the losers, generating cash to buy more promising stocks. You have used Excel for years in your work, and will use it as your analysis tool.

ASSIGNMENT 1: CREATING A SPREADSHEET FOR DECISION SUPPORT

In this assignment, you will produce a spreadsheet that models the problem and lets you construct a solution. Then, in Assignment 2, you will write a memorandum to your boss that explains your work and recommends a new portfolio. In addition, in Assignment 3, you may be asked to prepare an oral presentation of your analysis and recommendation.

A spreadsheet has been started. *The spreadsheet skeleton is available to you.* To access the spreadsheet skeleton, go to your Data files. Select Case 10 and then select **ATKINS.xlsx**.

Your spreadsheet has a worksheet named OWNED STOCKS that contains data on the 200 stocks in the portfolio. Figure 10-1 shows the top few rows of data.

	A	B	C	D	E	F	G	H
1	PORTFOLIO RECONSTRUCTION PROBLEM							
2								
3								
4	SYMBOL	SEGMENT	SHARES	COST	LAST	PRIOR	PE RATIO	EPS
5	MYOGO	RETAIL	12,500	$17.96	$23.50	$6.99	19	$1.24
6	STVK	RETAIL	5,000	$25.68	$25.90	$7.89	15	$1.73
7	NGSX	INDUSTRIAL	7,500	$18.85	$35.23	$11.70	23	$1.53
8	VPHMT	RETAIL	7,500	$22.05	$20.80	$6.95	27	$0.77
9	CNVRQ	INDUSTRIAL	5,000	$16.92	$14.10	$4.89	21	$0.67
10	ABPD	RETAIL	5,000	$21.44	$7.93	$2.76	17	$0.47

FIGURE 10-1 Stocks in portfolio

An explanation of the data values follows.

- SYMBOL: Common stocks are listed on the stock exchanges by a symbol, not by their company name. Each stock has its own unique symbol. Stocks are denoted here by their symbol only.
- SEGMENT: The stocks owned are categorized by their sector in the economy. The 200 stocks are in the retail, industrial, and financial sectors.
- SHARES: The number of shares owned of each company's common stock are shown.
- COST: The price paid for the shares owned is shown. For example, Mrs. Atkins owns 12,500 shares of stock MYOGO, and each cost $17.96 when bought for the portfolio.
- LAST: The most recent price of a share of the stock is shown. Continuing the MYOGO example, the most recent price of a share was $23.50.
- PRIOR: The 200 stocks have been bought over many years. No new stocks have been bought in a year. The price of a share of each stock a year ago is shown.
- PE RATIO: A stock's PE ratio is the ratio of its most recent stock price to its most recent earnings per share.
- EPS: EPS is a company's "earnings per share." This is the company's net income after taxes divided by the number of shares the company has sold to the public. EPS is an indicator of profitability—the higher the net income, the higher the EPS. Note that EPS is the denominator in the PE ratio, which is referred to in the previous cell.

Further explanation of the PE ratio and EPS might be helpful. Assume that a stock has an EPS of $5 and that a share of the stock could be bought on the stock exchange for $25. Its PE ratio would be $25 / $5 = 5. You would be paying five times the stock's earnings per share; in other words, its "multiple" is 5. Next, say that the price of a share of the stock goes up to $50 on the exchange. Its PE ratio is now $50 / $5 = 10. Thus, its multiple is now 10. Generally speaking, a lower multiple indicates less risk because the stock price has room to appreciate. Stocks with high multiples—high PE ratios—are usually considered riskier because of the fear that the price may have topped out and is about to decline. Thus, one conservative investment strategy would be to favor stocks with low PE ratios and avoid those with high PE ratios.

Reconstruction Strategy

Your reconstruction strategy will follow this sequence of steps:

1. Compute gains and losses on stocks individually and in total. This information will be needed when discussing portfolio changes with Mrs. Atkins.
2. Compute gains and losses for stocks by industrial sector (retail, industrial, and financial). Sector performance will be a factor later in deciding which stocks to keep and which to sell.
3. Compute the percentage price change for each stock. Some stock prices have gone up, some down. You will want to keep stocks that have gone up and think about selling those that have declined.
4. Identify the 50 stocks whose prices have declined the most. Segregate those as stocks to be sold.
5. Identify the 50 stocks whose prices have increased the most. Segregate those as stocks to keep in the portfolio.
6. Decide which of the remaining 100 stocks should be kept and which should be sold.

Next, you will work through the six steps.

Step 1: Compute gains and losses on stocks individually and in total

You should copy the OWNED STOCKS data to a new worksheet and name the sheet LOSSES. The OWNED STOCKS worksheet thus becomes your backup data in case something undesirable happens to your work.

Using the LOSSES data, you should compute each stock's current market value, its cost basis, and the gain or loss on the stock. Then compute these values for the 200 stocks in total. Start by pruning out unneeded columns, then by computing needed values. The first few rows of your worksheet should look like those in Figure 10-2.

	A	B	C	D	E	F	G	H	I
1	PORTFOLIO RECONSTRUCTION PROBLEM								
2						CURRENT			
3						MARKET		GAIN	
4	SYMBOL	SEGMENT	SHARES	COST	LAST	VALUE	COST	(LOSS)	NOTE
5	MYOGO	RETAIL	12,500	$17.96	$23.50	$293,750.00	$224,546.24	$69,203.76	GAIN
6	STVK	RETAIL	5,000	$25.68	$25.90	$129,500.00	$128,418.79	$1,081.21	GAIN
7	NGSX	INDUSTRIAL	7,500	$18.85	$35.23	$264,225.00	$141,408.54	$122,816.46	GAIN
8	VPHMT	RETAIL	7,500	$22.05	$20.80	$156,000.00	$165,369.89	-$9,369.89	LOSS
9	CNVRQ	INDUSTRIAL	5,000	$16.92	$14.10	$70,500.00	$84,591.25	-$14,091.25	LOSS
10	ABPD	RETAIL	5,000	$21.44	$7.93	$39,650.00	$107,218.61	-$67,568.61	LOSS

FIGURE 10-2   Computation of stock gains and losses

The Symbol, Segment, Shares, Cost, and Last values have been previously defined. An explanation of the other data values follows.

- CURRENT MARKET VALUE: This is a function of the number of shares and the current (LAST) share price.
- COST: This is a function of the number of shares and the cost of a share when the stock was bought.
- GAIN (LOSS): This is the stock's current market value less its cost basis.
- NOTE: If the market value exceeds its cost, then a "GAIN" on the stock has occurred; otherwise, a "LOSS" has occurred. Use an IF statement to record the proper note.

Next, compute these values for the 200 stocks in total, as shown in Figure 10-3:

	A	B	C	D	E	F	G	H	I
203	DALD	RETAIL	7,500	$13.30	$0.75	$5,625.00	$99,714.09	-$94,089.09	LOSS
204	NWACQN	INDUSTRIAL	7,500	$4.88	$0.71	$5,325.00	$36,569.18	-$31,244.18	LOSS
205									
206						TOTALS	$26,800,000.00	$42,300,000.00	-$15,500,000.00	
207						NUMBER OF LOSERS:			150
208						NUMBER OF GAINERS:			50

FIGURE 10-3   Total gains and losses

The totals shown here are arbitrary amounts for illustrative purposes. The column F total is the market value of all stocks, the column G total is the cost basis of all stocks, and the column H total is the total loss for all stocks. The number of losers and gainers can be computed using the Countif() function, using values in column I. The Countif() function is discussed in Tutorial E. (The numbers shown here are illustrative, but in your work you should see that there is an overall loss and that the Atkins portfolio is not in good shape.)

Step 2: Compute gains and losses for stocks by industrial sector

Copy the LOSSES sheet data to a new worksheet, renaming it as SECTORS. Make a data table out of the data range. (Note that when copying to another sheet, you can use Paste-Special Values to save space in memory. Also, a data table cannot have two identical column headings, so the heading for total cost should be changed.) Data table usage is discussed in Tutorial E.

Use the data table to compute total and average gains and losses for each of the three sectors. As data is computed you should manually fill out a summary schedule beneath the data table, as shown in Figure 10-4:

	A	B	C	D	E	F	G	H	I
206	SUMMARY DATA:								
207						CURRENT			
208						MARKET		GAIN	
209	SYMBOL	SEGMENT	SHARES	COST	LAST	VALUE	COST2	(LOSS)	NUMBER
210	FINANCIAL								
211	FINANCIAL--AVGS								
212	INDUSTRIAL								
213	INDUSTRIAL--AVGS								
214	RETAIL								
215	RETAIL--AVGS								

FIGURE 10-4   Summary of sector gains and losses

In the summary data, the NUMBER column is the number of stocks in each sector. The meanings of the other headings are the same as previously defined.

This data will become important when you get to Step 6. At that point, some stocks from the poorest-performing sector will be designated for sale. For example, assume that the average retail loss is greater than that of the other two sectors. At one point in Step 6, retail stocks would be designated for sale.

Step 3: Compute the percentage price change for each stock

Copy the OWNED STOCKS data to a new sheet and name it PCT CHANGE (you can leave out the EPS column). You want to compute the percentage change in price for each stock. At first you would think that the formula's logic should be:

(Last price – Cost) / Cost

However, not all the stocks were purchased at the same time, and you want to have the same starting point for all. Thus, your logic should be:

(Last – Prior) / Prior

Next, make a data table out of the data range. Sort the data in descending order using the PCT CHANGE column values. The first few rows of your data table should look like Figure 10-5:

	A	B	C	D	E	F	G	H
1								
2								
3								
4	PORTFOLIO RECONSTRUCTION PROBLEM							
5								
6								PCT
7	SYMBO	SEGMEN	SHARE	COST	LAST	PRIO	PE RATI	CHANG
8	MYOGO	RETAIL	12,500	$17.96	$23.50	$6.99	19	236.19%
9	STVK	RETAIL	5,000	$25.68	$25.90	$7.89	15	228.26%
10	NGSX	INDUSTRIAL	7,500	$18.85	$35.23	$11.70	23	201.11%
11	VPHMT	RETAIL	7,500	$22.05	$20.80	$6.95	27	199.28%

FIGURE 10-5   Percentage change in stock prices

You want to keep the 50 stocks that have the best percentage change and sell the 50 stocks that have the worst. The remaining 100 stocks will be further analyzed to see which should be kept and which sold.

Step 4: Identify the 50 stocks whose prices have declined the most

Make a new sheet and name it SELL. Use the Top 10 data table feature to identify the bottom 50 percent changes—the worst-performing stocks from the percent change standpoint. Copy the data for those stocks to the SELL sheet, the top part of which should look like Figure 10-6:

	A	B	C	D	E	F	G	H	I	J
1	STOCKS TO SELL									
2										
3										
4										
5							MARKET	COST		
6	SYMBO	SEGMENT	SHARE	COST	LAST	PRIOI	VALUE	BASIS	PE RATI	CHANG
7	DPHJ	FINANCIAL	7,500	$20.14	$2.76	$4.65	$20,700	$151,064	10	-40.65%
8	CRYOV	INDUSTRIAL	7,500	$13.13	$6.45	$10.87	$48,375	$98,453	20	-40.66%

FIGURE 10-6 Stocks to sell

Step 5: Identify the 50 stocks whose prices have increased the most

In the PCT CHANGE sheet, use the All feature to restore all data. Next, make a new sheet and name it KEEP. Use the Top 10 data table feature to identify the best 50 percent changes—the best-performing stocks from the percent change standpoint. Copy the data for those stocks to the KEEP sheet, the top part of which should look like Figure 10-7:

	A	B	C	D	E	F	G	H	I	J
1	STOCKS TO KEEP									
2										
3							MARKET	COST		
4	SYMBO	SEGMEN	SHARE	COST	LAST	PRIOI	VALUE	BASIS	PE RATI	CHANG
5	MYOGO	RETAIL	12,500	$17.96	$23.50	$6.99	$293,750	$224,546	19	236.19%
6	STVK	RETAIL	5,000	$25.68	$25.90	$7.89	$129,500	$128,419	15	228.26%

FIGURE 10-7 Stocks to keep

In the PCT CHANGE sheet, use the All feature to restore all data.

Step 6: Decide which of the remaining 100 stocks should be kept and which should be sold

Make a new sheet and name it WORK. Copy the PCT CHANGE data table to this sheet. Sort the data on the percent change. Delete the rows for the top 50 stocks and for the bottom 50 stocks, leaving a worksheet with 100 stocks that need further analysis. The purpose of the analysis is to decide which 50 stocks to keep and which 50 to designate for sale. You want to use the PE ratio and the sector data to make these designations.

Rank the remaining 100 stocks in the WORK sheet by their PE ratio. Of these stocks, you want to keep 50 and sell 50. Follow this six-step procedure:

1. Segregate for sale any stock whose PE ratio is over 24.
2. Segregate to keep the 25 lowest PE ratio stocks.
3. Segregate for sale stocks in the worst-performing sector.
4. Rank the remaining stocks by PE ratio.
5. Segregate to keep as many stocks as you need to get to 100 retained stocks.
6. Segregate for sale the remaining stocks.

Step 6-1: Segregate for sale any stock whose PE ratio is over 24   In the WORK sheet, sort the rows on PE ratio in descending order. You think that PE ratios over 24 are too risky. Copy rows for such stocks to the SELL sheet, then delete those rows from the WORK sheet.

Step 6-2: Segregate to keep 25 stocks with the lowest PE ratios In the WORK sheet, sort the rows on PE ratio in ascending order. You think that the lowest PE ratios indicate the least risk. Copy rows for the 25 lowest PE ratios to the KEEP sheet, then delete those rows from the WORK sheet.

Step 6-3: Segregate for sale stocks in the worst-performing sector Sort the remaining stocks by sector. Previously, the worst-performing sector had been identified. Copy the rows of data for the remaining stocks to the SELL sheet, then delete those rows from the WORK sheet.

Step 6-4: Rank the remaining stocks by PE ratio Sort remaining stocks by PE ratio.

Step 6-5: Segregate to keep as many stocks as you need to get to 100 retained stocks By inspection, determine how many stocks have been segregated to keep, and determine how many are needed to get to 100 kept stocks. Copy that many of the lowest PE ratios to the KEEP sheet, then delete those rows from the WORK sheet.

Step 6-6: Segregate for sale the remaining stocks The remaining stocks in the WORK sheet need to be sold. Copy that data to the SELL sheet.

ASSIGNMENT 2: USING THE SPREADSHEET FOR DECISION SUPPORT

You will now complete the case by (1) using the spreadsheet to gather data about the portfolio reconstruction, (2) documenting your findings and recommendation in a memorandum, and (3) if your instructor specifies, giving an oral presentation.

Assignment 2A: Using the Spreadsheet to Gather Data

You have built the spreadsheet to divide the portfolio into two groups of stocks: the 100 that should be kept and the 100 that should be sold. You can now create summary data about each group. Using data tables in the SELL and KEEP sheets, compute total market value, total cost, average PE ratio, and average percent price change for each 100-stock group. This gives you the summary data you need when talking with your boss, and then with Mrs. Atkins.

The difference between market and cost for the retained stocks, if positive, is a profit you will keep in the portfolio. If negative, the difference is a paper loss that you are willing to live with. You hope that the average PE ratio and average percent price changes for this group are acceptable to you, and that they support the case for keeping these stocks. You would need to evaluate that assumption, however.

The difference between market and cost for the stocks that will be sold is presumably a loss that must be accepted at this point to rationalize the portfolio's structure. Note that the group gives you the ability to generate cash that can be invested in more promising stocks. You hope that the average PE ratio and average percent price changes for this group are poor, and that they demonstrate the need for changes. You would need to evaluate that, however.

Summarize the data as shown in Figure 10-8.

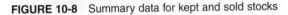

	Market Value	Cost Basis	Avg. PE Ratio	Avg. % Price Change
200 portfolio stocks				
100 retained stocks				
100 sold stocks				

FIGURE 10-8 Summary data for kept and sold stocks

Assignment 2B: Documenting Your Recommendation in a Memorandum

You are now in a position to summarize the results. When you finish with the spreadsheet, save the file for the last time, close it, and exit from Excel. You should write a memo to your boss summarizing the results of your work.

When writing the memo, observe the following requirements:

- Your memorandum should be set up as discussed in Tutorial E.
- Briefly state the purpose of the analysis and the analytical methods, but do not provide a lot of background. You can assume that your boss expects the memo and knows that you are working on the analysis.
- State the results of your analysis. Your memo should describe the current portfolio status and the status after selling half the stocks. The reader of the memo should understand your reasons for keeping and selling stocks. The reader also should understand what loss will be incurred in the sale and how much cash will be generated by the sale. Keep in mind that the memo will be the basis of what you tell Mrs. Atkins about her portfolio, and that the news will come as a surprise to her.

Support your analysis graphically. You should enter a table like that in Figure 10-8 in your Word document. The procedure for making a table is discussed in Tutorial E.

ASSIGNMENT 3: GIVING AN ORAL PRESENTATION

Assume that your boss is impressed by your analysis and thinks that other wealth managers would benefit from knowing more about it. He asks you to give a presentation explaining your methods and results. Prepare to explain your work and your findings to the group in 10 minutes or less. Use visual aids or handouts that you think are appropriate. Tutorial F has guidance on how to prepare and give an oral presentation.

DELIVERABLES

Assemble the following deliverables for your instructor:

1. Printout of your memo
2. Disk that contains your Word and Excel files. (Do not use a CD, which will be read-only.)

If there are other .xlsx files on your disk, write your instructor a note that lists the filenames pertinent to this case.

PART 5

INTEGRATION CASES: USING
ACCESS AND EXCEL

11

UNUSUAL COLLEGE GPA ANALYSIS

Decision Support Using Access and Excel

PREVIEW

Unusual College wants to know what factors (if any) strongly influence student GPA performance. You have been called in to use Access and Excel.

PREPARATION

- Review spreadsheet and database concepts discussed in class and/or in your textbook.
- Complete any exercises that your instructor assigns.
- Obtain the database file **COLLEGE.accdb** from your Data files in the Case 11 folder.
- Review any part of Tutorials A, B, C, or D that your instructor specifies, or refer to them as necessary.
- Review file-saving procedures for Windows programs. These are discussed in Tutorial C.
- Refer to Tutorials E and F as necessary for guidance on Excel data list processing and pivot tables, respectively.

BACKGROUND

Unusual College has some academic policies that are... unusual. The most unusual policy is that the college does not have academic probation. In other words, students who fall below a C average are not sent packing. Instead, they are promoted to the next academic year. Thus, all students make it to their senior year and then graduate into the real world, where they generally meet with unusually good success.

The college has recently hired a new dean of students. The dean is interested in knowing what factors seem to influence GPA performance the most. The dean thinks that the following factors might be influential:

- Gender
- Student's home state
- Whether the student is "Greek"—in other words, has joined a fraternity or sorority
- Student's academic year—freshman, sophomore, junior, or senior

The dean also thinks that factors may interact. Assume, for example, that GPA seems to be influenced by home state and by going Greek, but not by gender and academic year. Is there an interaction between Greeks and states? Is GPA influenced more for Greeks from only certain states, or is GPA better for Greeks from all the states?

The dean thinks it is plausible that males could have a higher GPA than females—or the reverse could be true. Unusual College draws students from three nearby states, code-named STATE1, STATE2, and STATE3 for this analysis. The college draws a good number of students from other states as well, and they are collectively code-named OTHER.

About half of Unusual's students join a fraternity or a sorority. The dean does not know whether going Greek helps a student's studies or hurts them. At the school that the dean just left, GPAs declined as academic years went on. In other words, seniors had lower GPAs on average than juniors, juniors had lower GPAs than sophomores, and so on. The dean assumes that the same trend exists at Unusual, but needs that assumption verified by data analysis.

The dean of students has called you in to analyze Unusual's GPA data. She has heard that you can use Access for database work and Excel for data analysis work, and that you can move data between the two programs in your analyses.

THE UNUSUAL COLLEGE STUDENT DATABASE

To continue, you must have the file **COLLEGE.accdb**, which you can find in your Data files in the Case 11 folder. The database has tables of data on the students and their academic performance. The database tables are discussed next, and then you will be told what is required in your analysis.

The STUDENT Table

The STUDENT table has data about the college's students. The table design is shown in Figure 11-1:

Field Name	Data Type
Student Number	Text
State	Text
Gender	Text
Greek	Text
Year	Text

FIGURE 11-1   STUDENT table design

Each student is given a student number. Names are omitted to protect student privacy. The State field indicates the student's home state: STATE1, STATE2, STATE3, or OTHER. Gender values can be F or M for Female and Male, respectively. The Greek field value can be Y or N for Yes and No, respectively. The Year field indicates the academic year just finished: FR, SO, JR, and SR, for freshman, sophomore, junior, and senior, respectively. There are 800 student records, 200 for each academic year. The first few records in the table are shown in Figure 11-2.

Student Nur ▾	State ▾	Gender ▾	Greek ▾	Year ▾
1001	STATE3	M	Y	SR
1002	STATE3	M	Y	SR
1003	STATE3	M	Y	SR

FIGURE 11-2   STUDENT data records

Data for student academic accomplishment is shown in the ACCOMPLISHMENT table, the design of which is shown in Figure 11-3:

Field Name	Data Type
Student Number	Text
Hours	Number
Credits	Number

FIGURE 11-3   ACCOMPLISHMENT table design

Again, each student is given a number. The Hours field denotes how many course hours the student has completed to date. The Credits field shows how many credits have been earned to date. The first few records in the table are shown in Figure 11-4.

Accomplishment		
Student Nur ▾	Hours ▾	Credits ▾
1001	120	462
1002	120	292
1003	120	381

FIGURE 11-4 ACCOMPLISHMENT data records

ASSIGNMENT 1: USING ACCESS AND EXCEL FOR DECISION SUPPORT

You will analyze the academic data in the following sequence of steps:

1. Make a data-gathering query in Access. The query output will be imported into Excel for analysis.
2. Make a blank spreadsheet called **COLLEGE.xlsx**. Import the query output into the file's **Sheet1**.
3. Copy data to a second sheet.
4. Use a pivot table to see if a relationship exists between GPA and year.
5. Use a pivot table to see if a relationship exists between GPA and home state.
6. Use a pivot table to see if a relationship exists between GPA and gender.
7. Use a pivot table to see if a relationship exists between GPA and Greek status.
8. Use a data table to analyze interactions between factors and develop other interesting statistics for the new dean.

Steps To Analyze Academic Data

Next, you will work through the following eight steps.

Step 1: Make a data-gathering query in Access

Make a query that outputs the student number, state, gender, Greek status, year, and GPA for each student. GPA will be a calculated field: credits divided by hours. Note that the query should have as many output rows as there are student records. Save the query as GPADATA. The first few rows of the output should look like Figure 11-5:

GPADATA					
Student Number ▾	State ▾	Gender ▾	Greek ▾	Year ▾	GPA ▾
1001	STATE3	M	Y	SR	3.850
1002	STATE3	M	Y	SR	2.433
1003	STATE3	M	Y	SR	3.175

FIGURE 11-5 GPADATA query output

When you finish with the query, close it and then close the database file.

Step 2: Import the query output into worksheet Sheet1

In Excel, open and save a new file called **COLLEGE.xlsx**. Import the GPADATA query output into **Sheet1**. Follow this procedure:

1. Click the **Data** tab.
2. In the Get External Data group, select **From Access**. In the Select Data Source dialog box, identify COLLEGE.accdb as the source file.
3. Select the **GPADATA** query from the Select Table dialog box and click **OK**.

4. The data will be brought into the worksheet as a table. To convert the table to a data range, click a cell in the table. The Design tab should activate. In the Tools group, select **Convert to Range** and then click **OK** in the Excel dialog box..

5. The cells may retain a formatted look, and you may want to change the formatting to a more standard black-and-white appearance. If so, click a blank cell outside the range and then select the format painter to capture the format of the empty cell. To select the format painter, click the **Home** tab, and in the **Clipboard** group, click **Format Painter**. Highlight the entire data range. When you release the cursor, the appearance will change to the standard look.

Step 3: Copy data to a separate sheet

You should rename Sheet1 to BACKUP. Then copy the data to another worksheet, renaming that sheet as DATA. To copy data, highlight the data range. Then activate the Home tab and select Copy in the Clipboard group. Move to the target sheet and click in the upper-left cell of the desired range. In the Clipboard group, select Paste, and then select Paste again from the drop-down menu.

You will now work with the DATA sheet, keeping the BACKUP sheet in case something happens to your data.

Step 4: Use a pivot table to see if a relationship exists between GPA and year

On average, do freshmen do better than sophomores, who do better than juniors, who do better than seniors? If that were so, the new dean of students would think about putting very poor students on probation— why retain poor students if they get worse with time? Or does progress run in the other direction? If that were the case, the new dean would want to keep the existing policy—why jettison struggling students if they improve with time? Or is there no real difference by year?

To answer these questions, make a pivot table that relates academic year with average GPA. Save the pivot table to a separate sheet, which you should name BY YEAR. (Pivot tables are discussed in Tutorial E.) In a column next to the pivot table, by Excel formula show the difference between average freshman GPA and average sophomore GPA, between average sophomore GPA and average junior GPA, and between average junior GPA and average senior GPA.

The dean tells you that if each year-to-year difference increases (or decreases) and if each change is at least .03, you can conclude that there is a relationship between academic year and average GPA. Otherwise, you can conclude that there is no relationship between the two.

If there is a relationship, you should make a column chart that illustrates it. If there is no relationship, no chart is needed. Put the chart in a separate sheet, which you should name BY YEAR CHART. (Charting in Excel is discussed in Tutorial F.)

Step 5: Use a pivot table to see if a relationship exists between GPA and home state

Do students from particular states outperform the average student? For example, do students from STATE1 do better than the average student? If so, the new dean of students would think about putting more recruiting effort into states whose students do much better than the average. If GPA does not vary much by state, then the new dean would keep the existing recruiting policy, which is to recruit evenly in the region.

To answer this question, make a pivot table that relates state with average GPA. Put the pivot table in a separate sheet named BY STATE. In a column next to the pivot table, by Excel formula show the difference between the state's average and the overall average for all states.

The dean tells you that if a state's average GPA is .10 or more better than the overall average, she would consider beefing up recruiting in that state. Are there any such states? If so, make a column chart that shows the state averages. Put the chart in a separate sheet, which you should name BY STATE CHART.

Step 6: Use a pivot table to see if a relationship exists between GPA and gender

Does one gender do better than the other on average? In other words, is the average GPA for women much better than for men, or vice versa? If that were so, the new dean of students might change some educational practices on campus. She might think about changing recruiting practices to recruit more of the gender that does better, if it is legal to do so.

To answer these questions, make a pivot table that relates gender with average GPA. Put the pivot table in a separate sheet named BY GENDER. In a column next to the pivot table, by Excel formula show the difference between gender average and the overall average for both genders.

The dean tells you that if one gender's average GPA is .10 better (or worse) than the other's, you can conclude that there is a relationship between gender and average GPA. Otherwise, you can conclude that there is no relationship.

If there is a relationship, you should make a column chart that illustrates it. Put the chart in a separate sheet, which you should name BY GEN CHART.

Step 7: Use a pivot table to see if a relationship exists between GPA and Greek status

On average, do students who join a fraternity or sorority do better than non-Greek students? If that were so, the new dean of students would think about recognizing and rewarding Greek organizations in some way. If Greeks do much worse, the new dean would want to encourage the Greeks to study a little harder.

To answer this question, make a pivot table that relates Greek status with average GPA. Put the pivot table in a separate sheet named BY GREEK. In a column next to the pivot table, by Excel formula show the difference between Greek status average and the overall student average.

The dean tells you that if average GPA differs by .10 or more, you can conclude that a relationship exists between Greek status and average GPA. Is there a relationship? If so, make a column chart that shows the two averages. Put the chart in a separate sheet, which you should name BY GREEK CHART.

Step 8: Use a data table to analyze interactions between factors and develop other interesting statistics for the new dean

Copy data values (including column headings) from the DATA sheet to another worksheet. Rename that sheet INTERACTION.

Delete data columns for factors that have no relationship to average GPA. Make a data table from the remaining data values. (Tutorial E covers making and using data tables.) For illustrative purposes only, assume you have shown that relationships exist between state and GPA and between Greek and GPA. The top part of your data table might look like Figure 11-6:

	A	B	C	D
1	Student Number	State	Greek	GPA
2	1001	STATE3	Y	3.850
3	1002	STATE3	Y	2.433
4	1003	STATE3	Y	3.175

FIGURE 11-6   Top part of data table

You want to see how the state and Greek factors relate. You would use the data table to compute average GPA for Greek and non-Greek students in each state. Beneath the data table, you would summarize results in the format shown in Figure 11-7. You would manually enter data as it is developed by using the data table.

Continuing the example, Figure 11-8 shows sample (and arbitrary) data.

Here is how to interpret this sample data: There are 92 non-Greek students from OTHER states. Their average GPA was 2.867. There are 800 students in total and their average GPA is 2.914. The difference between the average GPA for non-Greeks from other states and the average GPA for all students is .047.

In the sample data, note that Greek students from STATE1 had the highest average GPA. Non-Greek students from STATE3 had the worst average GPA.

The conclusion should be verified in the following way: Use the data table to identify the students with the top 100 GPAs. Of those 100, how many are in the STATE1-Greek group? There should be many more than one-eighth of the 100. Use the data table to identify the students with the bottom 100 GPAs. Of those 100, how many (continuing the arbitrary example) are in the STATE3-non-Greek group? There should be many more than one-eighth of the 100.

The data table should be used to verify averages seen in the pivot tables. In the example, the average GPA for Greeks and non-Greeks can be verified, and the average GPA for the states and for all students can be verified.

	A	B	C	D
805	**AVERAGES**	**NUMBER**	**GPA**	**DIFF**
806	OTHER-NO			
807	OTHER-YES			
808	STATE1-NO			
809	STATE1-YES			
810	STATE2-NO			
811	STATE2-YES			
812	STATE3-NO			
813	STATE3-YES			
814				
815	YES-ALL			
816	NO-ALL			
817				
818	STATE1-ALL			
819	STATE2-ALL			
820	STATE3-ALL			
821	OTHER-ALL			
822				
823	ALL STUDENTS			

FIGURE 11-7   Form of summary data beneath data table

	A	B	C	D	E	F	G
805	**AVERAGES**	**NUMBER**	**GPA**	**DIFF**			
806	OTHER-NO	92	2.867	-0.047			
807	OTHER-YES	105	3.009	0.095			
808	STATE1-NO	102	2.812	-0.102			
809	STATE1-YES	111	3.315	0.402		STATE1-YES IS BEST	
810	STATE2-NO	111	2.666	-0.248			
811	STATE2-YES	87	3.100	0.187			
812	STATE3-NO	86	2.555	-0.359		STATE3-NO IS WORST	
813	STATE3-YES	106	2.966	0.053			
814							
815	YES-ALL	409	3.151	0.238		GREEKS DO BETTER	
816	NO-ALL	391	2.676	-0.238		NONGREEKS DO WORSE	
817							
818	STATE1-ALL	213	3.100	0.187		STATE1 BETTER THAN AVG	
819	STATE2-ALL	198	2.766	-0.148		STATE2 WORSE	
820	STATE3-ALL	192	2.788	-0.126		STATE3 WORSE	
821	OTHER-ALL	197	3.000	0.087		OTHER NA	
822							
823	ALL STUDENTS	800	2.914	0.000		ALL AVG = 2.914	

FIGURE 11-8   Summary data beneath data table

The data table analysis would continue as follows. STATE1-Greek shows the strongest GPA interaction; in other words, Greek students from STATE1 have the best average GPA. Does their GPA appear to be the best because STATE1 students are so good or because Greek students are so good, or is it a combination of the two factors? Here, it appears to be a combination of factors because the average GPA for STATE1 is 3.1 and the average GPA for all Greeks is 3.151. The two averages are about the same, and so one factor would not dominate the other.

The dean would like you to perform two final data table tasks. Use the data table to identify the senior valedictorian; in other words, the student with the highest GPA. Also, the U.S. military academies acknowledge the graduate with the worst GPA, and the dean would like to adopt that practice at Unusual. Thus, you should identify the senior with the lowest GPA.

ASSIGNMENT 2: USING THE SPREADSHEET FOR DECISION SUPPORT

You are now in a position to summarize the results. When you finish with the spreadsheet, save the file for the last time, close it, and exit from Excel. You should write a memo to the new dean summarizing the results of your analysis. Your memo should answer the questions previously raised by the dean:

1. Is there a relationship between GPA and year?
2. Is there a relationship between GPA and home state?
3. Is there a relationship between GPA and gender?
4. Is there a relationship between GPA and Greek status?
5. What interactions exist between significant factors?
6. Who is the senior valedictorian?
7. Who is the senior bringing up the rear?

When writing the memo, observe the following requirements:

- Your memorandum should be set up as discussed in Tutorial E.
- Briefly state the purpose of the analysis and the analytical methods, but do not provide a lot of background. You can assume that the dean expects the memo and knows that you are working on the analysis.
- State the results of your analysis—answer the dean's questions. What are the policy implications for each question?

Support your analysis by including a table that summarizes important results. The procedure for inserting a table in a Word document is described in Tutorial E.

Your table should look like Figure 11-9.

	Relationship to GPA? (Y/N)	Policy implication to consider
Year		
Home state		
Gender		
Greek status		

FIGURE 11-9 Form for table in memorandum

The table should summarize findings on four of the dean's questions. You should discuss interesting interactions in a separate paragraph. Identify the valedictorian and the lowest-ranking graduate in a final paragraph.

ASSIGNMENT 3: GIVING AN ORAL PRESENTATION

Assume that the new dean is impressed by your analysis and thinks that Unusual College's vice presidents need to know more about it. She asks you to give a presentation explaining your methods and results. Prepare to explain your work and your findings to the group in 10 minutes or less. Use visual aids or handouts that you think are appropriate. Of course, you would use any Excel charts created during your analysis. Tutorial F has guidance on how to prepare and give an oral presentation.

DELIVERABLES

Assemble the following deliverables for your instructor:

1. Printout of your memo and Excel charts
2. Disk that contains your Word, Excel, and Access files. (Do not use a CD, which will be read-only.)

If there are other .xlsx files or .accdb files on your disk, write your instructor a note that lists the filenames pertinent to this case.

THE DVD RENTAL PREDICTION

Designing a Relational Database to Create Tables, Queries, and Decision Support Using Access and Excel

PREVIEW

In this case, you'll design a relational database for a company that rents DVDs by mail. Once the database is designed, you'll create the tables and populate them with data. You need some data from the Access database to analyze in Excel. You'll create a query to import data into Excel, use the least squares method to predict what DVDs a given customer will likely rent, and then export the information back to Access for future business use.

PREPARATION

- Review spreadsheet and database concepts discussed in class and/or in your textbook.
- Complete any part of Database Design Tutorial A that your instructor assigns.
- Complete any part of Tutorials B, C, or D that your instructor assigns, or refer to the tutorials as necessary.
- Refer to Tutorials E and F as necessary.

BACKGROUND

You have just landed a job with a new company that rents DVDs by mail to customers. This new company, DVDs at Your Door (DYD), competes with other companies for an ever-expanding market. The business is just getting off the ground, and you have been hired this summer to create prototype systems for DYD to use as it ramps up to become a full-fledged company. Your job will not only entail designing and creating the database, you will analyze the data for future marketing purposes. Companies that rent films want to know how to predict what films customers want to see. This helps for marketing purposes and for stocking the inventory at proper levels. DYD feels that with its new ideas for predicting what DVDs customers will rent, it can snatch customers from the competition and earn lots of money. Your expertise in Access and Excel is a perfect fit for the company.

Before you begin to design this database, you must keep several parameters in mind. Obviously you must record customer details such as name, address, phone number, e-mail address, and billing information. You will also record which DVDs customers rent and their rating for each one. (Customers are asked to rate the films when they return them.) Your DVD inventory is huge, and information about these DVDs also must be recorded. Along with information such as title, year, genre, rating, and awards, you need to record the rating that DYD has given to each film. You are not allowed to see this proprietary formula; it takes into account many parameters for the film. DYD gives each film a rating to be able to correlate the customers' ratings and predict what they will rent next.

After your database is complete and populated with data, you can begin the analysis portion of the data. DYD wants to be able to export a given customer's rental information: their DVDs rented, the attributes of those DVDs, and the customer rating of the DVDs. With this information in Excel, you can run a least squares fit of the data. The least squares fit is a mathematical procedure that finds the best-fitting curve to a set of points. From this fit, you can make future predictions about DVD rentals. Then, using attributes from DVDs that a particular customer has never watched (also exported from Access), you will predict whether the customer might be interested in renting the film.

The list of potential rental films will then be imported back to Access, where the system can query the database and make recommendations for the customer's next choices.

The owners of DVDs at Your Door realize that the system you are creating is simply a model or prototype of the final system. They want to attract a huge number of customers, and they understand that your Access database and analysis of one customer will not be sufficiently robust. However, your job is important to lay the groundwork for the new company's future information systems.

ASSIGNMENT 1: CREATING THE DATABASE DESIGN

In this assignment, you will design your database tables on paper using a word-processing program. Pay close attention to the tables' logic and structure. Do not start your Access code (Assignment 2) before getting feedback from your instructor on Assignment 1. Keep in mind that you need to look at the requirements in Assignment 2 to design your fields and tables properly. It's good programming practice to look at the required outputs before designing your database. When designing the database, observe the following guidelines:

- First, determine the tables you'll need by listing on paper the name of each table and the fields it should contain. Avoid data redundancy. Do not create a field if it could be created by a "calculated field" in a query.
- You'll need at least one transaction table. Avoid duplicating data.
- Document your tables by using the Table facility of your word processor. Your word-processed tables should resemble the format of the table in Figure 12-1.
- You must mark the appropriate key field(s). You can designate a key field using an asterisk (*) next to the field name. Keep in mind that some tables might need a compound primary key to uniquely identify a record within a table.
- Print the database design.

Table Name	
Field Name	Data Type (text, numeric, currency, etc.)
...	...
...	...

FIGURE 12-1 Table design

NOTE

Have your design approved before beginning Assignment 2; otherwise, you may need to redo Assignment 2.

ASSIGNMENT 2: CREATING THE DATABASE AND MAKING QUERIES AND A REPORT; USING EXCEL FOR DECISION SUPPORT; IMPORTING AND EXPORTING DATA

In this assignment, you will first create database tables in Access and populate them with data. Next, you will create two queries to export to Excel. You'll analyze the data in Excel and return it to Access.

Assignment 2A: Creating Tables in Access

In this part of the assignment, you will create your tables in Access. Use the following guidelines:

- Type records into the tables, using yourself as the only customer.
- Include in your record 10 films that you have seen recently. Make sure you rate each film on a scale from 1 to 100.
- In addition, make sure that DYD rents at least 20 films, each with a DYD rating. Assume there is only one DVD per film (not multiple copies).
- Record both genre and film rating using a number system. For example, G-rated films are "1", PG-rated films are "2", and so on.

- Appropriately limit the size of the text fields; for example, a zip code does not need the default setting of 255 characters.
- Print all tables, if required.

Assignment 2B: Creating Queries

You must create two queries and export them to Excel, as outlined in the Background section of this case.

Query 1

Create a query called Customer's Ratings that lists the DVD ID, Customer Score, DYD score, Genre, and Rating for all the films you have watched. Export the query output to an Excel worksheet. The data, although different, should resemble Figure 12-2.

DVD ID	Customer Score	DYD score	Genre	Rating
101	89	56	1	5
102	80	55	1	5
103	90	54	1	4
104	92	54	1	4
105	55	89	3	2
106	96	51	1	4
107	98	52	1	4
108	87	67	3	5
109	23	88	2	1
110	99	49	1	2

FIGURE 12-2   Customer's Ratings query output exported to Excel

Query 2

Create a query called Films Not Watched. List the DVD ID, DYD score, Genre, and Rating of the films you did NOT watch. Export the output to an Excel worksheet. Your data will differ, but your output should resemble Figure 12-3.

DVD ID	DYD score	Genre	Rating
201	50	1	4
202	90	3	2
203	45	1	2
204	78	3	4
205	23	1	2
206	98	4	1
207	76	3	2
208	91	3	5
209	50	1	2
210	78	2	1

FIGURE 12-3   Films Not Watched query output exported to Excel

Assignment 2C: Analyzing and Exporting Data

You need to analyze the data and export it back to Access.

Least Squares Analysis

Using the worksheets that you imported in Excel, use the =TREND function to create a least squares fit to the customer data and predict the score of the customer on DVDs not seen. Using the =IF function, assume that any film with a score over 88 would be recommended and that films under that score would not be recommended. Your analysis will yield results that resemble the data in Figure 12-4.

DVD ID	DVD score	Genre	Rating	Predicted Rating	
201	50	1	4	98.080276	Recommend
202	90	3	2	42.794967	Don't Recommend
203	45	1	2	109.77579	Recommend
204	78	3	4	68.828207	Don't Recommend
205	23	1	2	158.60123	Recommend
206	98	4	1	41.784373	Don't Recommend
207	76	3	2	73.865704	Don't Recommend
208	91	3	5	39.677398	Don't Recommend
209	50	1	2	98.679096	Recommend
210	78	2	1	53.281734	Don't Recommend

FIGURE 12-4   Analysis and recommendations in Excel

Export to Access

Prepare the data to be exported into a new table in Access. Move the important information to a separate sheet complete with column headings, such as the table shown in Figure 12-5. Export the data and create a table called Predictions. Data will differ, but your new table in Access should resemble the one in Figure 12-5.

Predictions			
DVD ID	Predicted Rating	Recommend?	Add New Field
201	98.0802759322018	Recommend	
202	42.794966666137	Don't Recommend	
203	109.775788061551	Recommend	
204	68.8282074621547	Don't Recommend	
205	158.601233023791	Recommend	
206	41.784372921599	Don't Recommend	
207	73.8657043693805	Don't Recommend	
208	39.6773981200473	Don't Recommend	
209	98.6790960246786	Recommend	
210	53.2817341326499	Don't Recommend	

FIGURE 12-5   Exported data in Access

ASSIGNMENT 3: MAKING A PRESENTATION

Create a presentation for DVDs at Your Door that includes information about your database design. Demonstrate how your prediction works and suggest further predictive indicators such as award winners and other categories. Include time for discussing how you will expand your analysis to include all customers' data. Your presentation should take less than 15 minutes, including a brief question-and-answer period.

DELIVERABLES

Assemble the following deliverables for your instructor, either electronically or in printed form:

1. Word-processed design of tables
2. Tables created in Access
3. Query 1: Customer's Ratings
4. Query 2: Films Not Watched
5. Excel Worksheet
6. Access Table: Predictions
7. Presentation materials
8. Any other required tutorial printouts or tutorial diskette, CD, or USB memory stick

Staple all pages together. Put your name and class number at the top of the page. Make sure your diskette, CD, or USB memory stick is labeled, if required.

PART **6**

ADVANCED SKILLS USING
EXCEL

TUTORIAL E

GUIDANCE FOR EXCEL CASES

The Excel cases in this book require the student to write a memorandum that includes a table. Guidelines for preparing a memorandum in Microsoft Word and instructions for entering a table in a Word document are listed to begin this tutorial. Also, some of the cases in this casebook require the use of advanced Excel techniques. Those techniques are explained in this tutorial rather than in the cases themselves:

- Using data tables
- Using pivot tables
- Using built-in functions

You can refer to Sheet 1 of TutEData.xlsx when reading about data tables. Refer to Sheet 2 when reading about pivot tables.

PREPARING A MEMORANDUM IN WORD

A business memorandum should include proper headings, such as TO/ FROM/ DATE/ SUBJECT. If you want to use a Word memorandum template, follow these steps:

1. In Microsoft Word, click the **Office Button**.
2. Click **New**.
3. Click the **Memos** template in the **Templates** section.
4. Choose the contemporary design memo.
5. Click **OK**.

The first time you do this, you may need to click Download to install the template.

ENTERING A TABLE INTO A WORD DOCUMENT

Enter a table into a Word document using the following procedure:

1. Click the cursor where you want the table to appear in the document.
2. In the **Insert** group, select the **Table** drop-down menu.
3. Select **Insert Table**.
4. Choose the number of rows and columns.
5. Click **OK**.

DATA TABLES

An Excel data table is a contiguous range of data that has been designated as a table. Once you make this designation, the table gains certain properties that are useful for data analysis. (*Note:* In previous versions of Excel, data tables were called *data lists*.) Suppose you have a list of runners who have run a race, as shown in Figure E-1.

	A	B	C	D	E	F
1	**RUNNER#**	**LAST**	**FIRST**	**AGE**	**GENDER**	**TIME (MIN)**
2	100	HARRIS	JANE	O	F	70
3	101	HILL	GLENN	Y	M	70
4	102	GARCIA	PEDRO	M	M	85
5	103	HILBERT	DORIS	M	F	90
6	104	DOAKS	SALLY	Y	F	94
7	105	JONES	SUE	Y	F	95
8	106	SMITH	PETE	M	M	100
9	107	DOE	JANE	O	F	100
10	108	BRADY	PETE	O	M	100
11	109	BRADY	JOE	O	M	120
12	110	HEEBER	SALLY	M	F	125
13	111	DOLTZ	HAL	O	M	130
14	112	PEEBLES	AL	Y	M	63

FIGURE E-1   Data table example

To turn the information into a data table (list), you highlight the data range, including headers, and select the Insert tab. Then, in the Tables group, click **Table**. You see the Create Table window, as shown in Figure E-2.

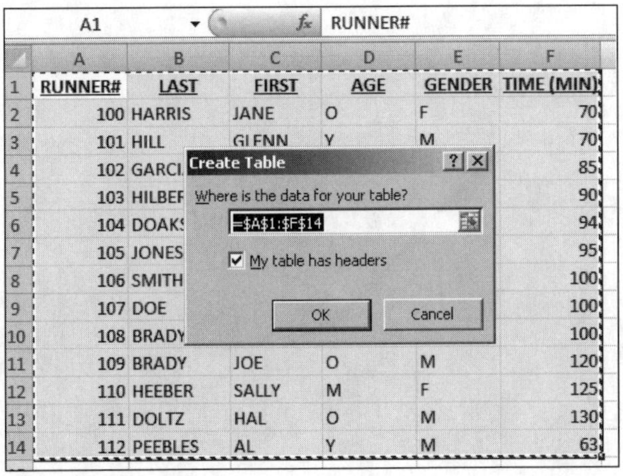

FIGURE E-2   Create Table window

When you click **OK**, the data range appears as a table. In the Design tab, select **Total Row** to add a totals row to the data table. You also can select **Table Styles – Light** to get rid of the contrasting color in the table's rows. Figure E-3 shows the results.

	A	B	C	D	E	F
1	RUNNER#	LAST	FIRST	AGE	GENDER	TIME (MIN)
2	100	HARRIS	JANE	O	F	70
3	101	HILL	GLENN	Y	M	70
4	102	GARCIA	PEDRO	M	M	85
5	103	HILBERT	DORIS	M	F	90
6	104	DOAKS	SALLY	Y	F	94
7	105	JONES	SUE	Y	F	95
8	106	SMITH	PETE	M	M	100
9	107	DOE	JANE	O	F	100
10	108	BRADY	PETE	O	M	100
11	109	BRADY	JOE	O	M	120
12	110	HEEBER	SALLY	M	F	125
13	111	DOLTZ	HAL	O	M	130
14	112	PEEBLES	AL	Y	M	63
15	Total					1242

FIGURE E-3   Data table example

The headers have acquired drop-down menu tabs, as you can see in Figure E-3.

You can sort the data table records by any field. Perhaps you want to sort by Times. If so, click the drop-down menu in the **TIME (MIN)** header and select **Sort – Smallest to Largest**. You get the results shown in Figure E-4.

	A	B	C	D	E	F
1	RUNNER	LAST	FIRST	AGE	GENDE	TIME (MIN
2	112	PEEBLES	AL	Y	M	63
3	100	HARRIS	JANE	O	F	70
4	101	HILL	GLENN	Y	M	70
5	102	GARCIA	PEDRO	M	M	85
6	103	HILBERT	DORIS	M	F	90
7	104	DOAKS	SALLY	Y	F	94
8	105	JONES	SUE	Y	F	95
9	106	SMITH	PETE	M	M	100
10	107	DOE	JANE	O	F	100
11	108	BRADY	PETE	O	M	100
12	109	BRADY	JOE	O	M	120
13	110	HEEBER	SALLY	M	F	125
14	111	DOLTZ	HAL	O	M	130
15	Total					1242

FIGURE E-4   Sorting list by drop-down menu

You can see that Peebles had the best time and Doltz had the worst time. You also can sort from Largest to Smallest.

In addition, you can sort by more than one criterion. Assume you want to sort first by Gender and then by Time (within Gender). You first sort from Smallest to Largest in Gender. Then you again activate the Gender drop-down tab and select **Sort By Color – Custom Sort**. In the Sort window that appears, click **Add Level** and choose **Time** as the next criterion. See Figure E-5.

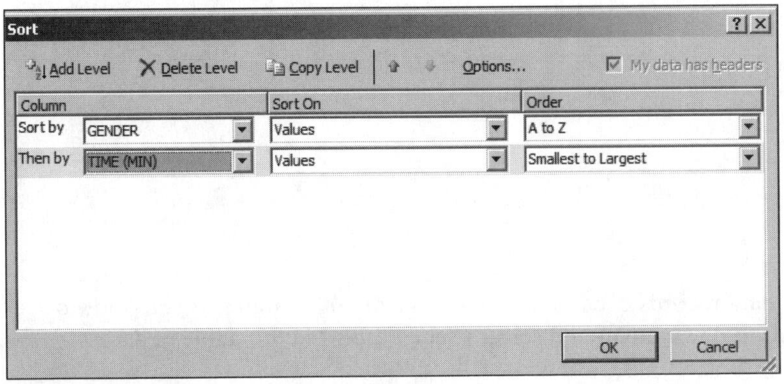

FIGURE E-5   Sorting on multiple criteria

Click **OK** to get the results shown in Figure E-6.

	A	B	C	D	E	F
1	RUNNER	LAST	FIRST	AGE	GENDE	TIME (MIN
2	100	HARRIS	JANE	O	F	70
3	103	HILBERT	DORIS	M	F	90
4	104	DOAKS	SALLY	Y	F	94
5	105	JONES	SUE	Y	F	95
6	107	DOE	JANE	O	F	100
7	110	HEEBER	SALLY	M	F	125
8	112	PEEBLES	AL	Y	M	63
9	101	HILL	GLENN	Y	M	70
10	102	GARCIA	PEDRO	M	M	85
11	106	SMITH	PETE	M	M	100
12	108	BRADY	PETE	O	M	100
13	109	BRADY	JOE	O	M	120
14	111	DOLTZ	HAL	O	M	130
15	Total					1242

FIGURE E-6   Sorting by Gender and Time (within Gender)

You can see that Harris had the best female time and that Peebles had the best male time.

Perhaps you want to see the Top n listings for some attribute; for example, you may want to see the top five runners' times. Select the Time column's drop-down menu and select **Number Filters**. You get another menu, in which you select **Top 10**. The Top 10 AutoFilter window appears, as shown in Figure E-7.

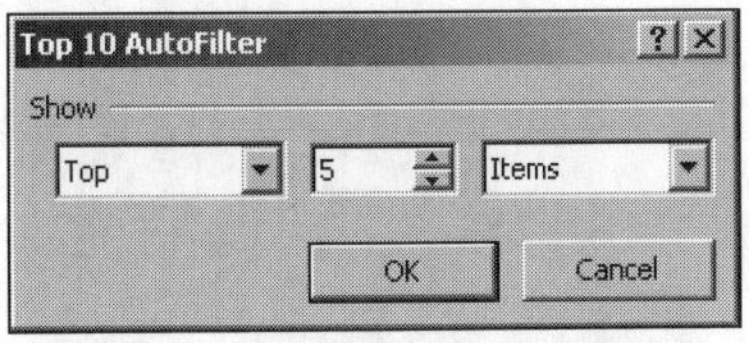

FIGURE E-7   Top 10 AutoFilter

This window lets you specify the number of values you want. In Figure E-7, five values were specified. Click **OK** to get the results shown in Figure E-8.

	A	B	C	D	E	F
1	RUNNER ▾	LAST ▾	FIRST ▾	AGE ▾	GENDE ▾	TIME (MIN ▾
6	107 DOE		JANE	O	F	100
7	110 HEEBER		SALLY	M	F	125
11	106 SMITH		PETE	M	M	100
12	108 BRADY		PETE	O	M	100
13	109 BRADY		JOE	O	M	120
14	111 DOLTZ		HAL	O	M	130
15	Total					675

FIGURE E-8   Top 5 times

The output contains more than five data records because there are ties at 100 minutes. If you want to see all of the records again, click the Time drop-down menu and select **Clear Filter**. The full table of data reappears, as shown in Figure E-9.

	A	B	C	D	E	F
1	RUNNER ▾	LAST ▾	FIRST ▾	AGE ▾	GENDE ▾	TIME (MIN ▾
2	100 HARRIS		JANE	O	F	70
3	103 HILBERT		DORIS	M	F	90
4	104 DOAKS		SALLY	Y	F	94
5	105 JONES		SUE	Y	F	95
6	107 DOE		JANE	O	F	100
7	110 HEEBER		SALLY	M	F	125
8	112 PEEBLES		AL	Y	M	63
9	101 HILL		GLENN	Y	M	70
10	102 GARCIA		PEDRO	M	M	85
11	106 SMITH		PETE	M	M	100
12	108 BRADY		PETE	O	M	100
13	109 BRADY		JOE	O	M	120
14	111 DOLTZ		HAL	O	M	130
15	Total					1242

FIGURE E-9   Restoring all data to screen

Each of the cells in the Total row has a drop-down menu. The menu choices are statistical operations that you can perform on the totals—for example, you can take a sum, take an average, take a minimum or maximum, count the number of records, and so on. Assume the Time drop-down menu was selected, as shown in Figure E-10. Note that the Sum operator is highlighted by default.

	A	B	C	D	E	F
1	RUNNER	LAST	FIRST	AGE	GENDE	TIME (MIN
2	100 HARRIS	JANE		O	F	70
3	103 HILBERT	DORIS		M	F	90
4	104 DOAKS	SALLY		Y	F	94
5	105 JONES	SUE		Y	F	95
6	107 DOE	JANE		O	F	100
7	110 HEEBER	SALLY		M	F	125
8	112 PEEBLES	AL		Y	M	63
9	101 HILL	GLENN		Y	M	70
10	102 GARCIA	PEDRO		M	M	85
11	106 SMITH	PETE		M	M	100
12	108 BRADY	PETE		O	M	100
13	109 BRADY	JOE		O	M	120
14	111 DOLTZ	HAL		O	M	130
15	Total					1242
16						None
17						Average
18						Count
19						Count Numbers
20						Max
21						Min
22						Sum
						StdDev
						Var
						More Functions…

FIGURE E-10   Selecting Time drop-down menu in Total row

By changing from Sum to the Average operator, you find that the average time for all runners was 95.5 minutes, as shown in Figure E-11.

	A	B	C	D	E	F
1	RUNNER	LAST	FIRST	AGE	GENDE	TIME (MIN
2	100 HARRIS	JANE		O	F	70
3	103 HILBERT	DORIS		M	F	90
4	104 DOAKS	SALLY		Y	F	94
5	105 JONES	SUE		Y	F	95
6	107 DOE	JANE		O	F	100
7	110 HEEBER	SALLY		M	F	125
8	112 PEEBLES	AL		Y	M	63
9	101 HILL	GLENN		Y	M	70
10	102 GARCIA	PEDRO		M	M	85
11	106 SMITH	PETE		M	M	100
12	108 BRADY	PETE		O	M	100
13	109 BRADY	JOE		O	M	120
14	111 DOLTZ	HAL		O	M	130
15	Total					95.53846154

FIGURE E-11   Average running time shown in Total row

PIVOT TABLES

Suppose you have data for a company's sales transactions by month, by salesperson, and by amount for each product type. You would like to display each salesperson's total sales by type of product sold and by month. You can use a pivot table in Excel to tabulate that summary data. A pivot table is built around one or more dimensions and thus can summarize large amounts of data.

Figure E-12 shows total sales cross-tabulated by salesperson and by month. The following steps explain how to create a pivot chart from the data.

	A	B	C	D	E
1	**Name**	**Product**	**January**	**February**	**March**
2	Jones	Product1	30,000	35,000	40,000
3	Jones	Product2	33,000	34,000	45,000
4	Jones	Product3	24,000	30,000	42,000
5	Smith	Product1	40,000	38,000	36,000
6	Smith	Product2	41,000	37,000	38,000
7	Smith	Product3	39,000	50,000	33,000
8	Bonds	Product1	25,000	26,000	25,000
9	Bonds	Product2	22,000	25,000	24,000
10	Bonds	Product3	19,000	20,000	19,000
11	Ruth	Product1	44,000	42,000	33,000
12	Ruth	Product2	45,000	40,000	30,000
13	Ruth	Product3	50,000	52,000	35,000

FIGURE E-12   Excel spreadsheet data

You can create pivot tables and many other kinds of tables with the Excel PivotTable tool. To create a pivot table from the data in Figure E-12, follow these steps:

1. Starting in the spreadsheet in Figure E-12, go to the Insert tab. In the Tables group, choose PivotTable. You see the screen shown in Figure E-13.

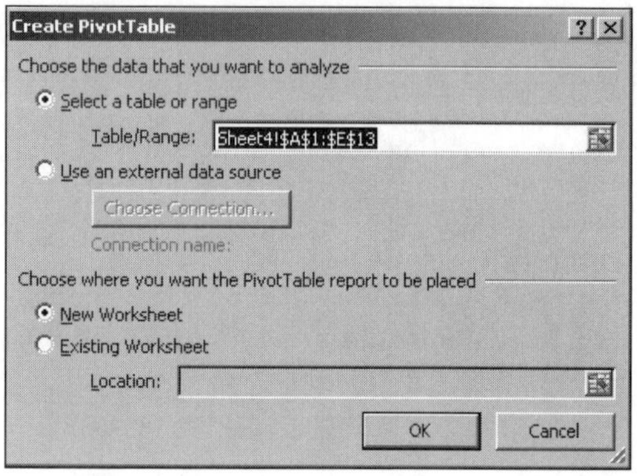

FIGURE E-13   Creating a pivot table

2. Make sure New Worksheet is checked under "Choose where you want the PivotTable report to be placed." Click **OK**. The screen shown in Figure E-14 appears. If it does not, right-click in a cell in the pivot table area. Select Pivot Table Options from the menu. Click the Display tab and then check the Classic Layout.

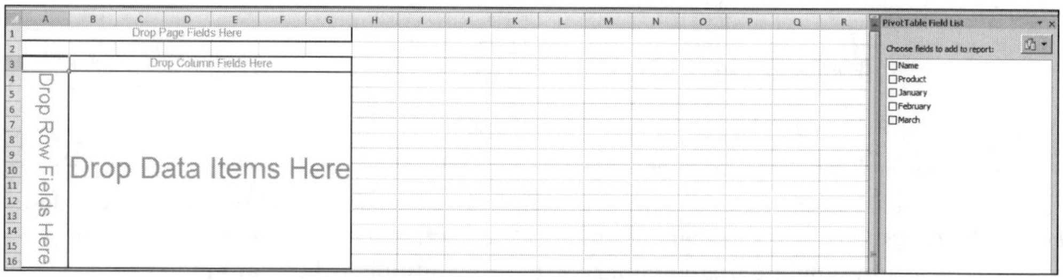

FIGURE E-14   PivotTable design screen

The data range's column headings are shown in the PivotTable Field List on the right side of the screen. From there, you can click and drag column headings into the Row, Column, and Data areas that appear in the spreadsheet.

3. If you want to see the total sales by product for each salesperson, drag the Name field to the Drop Column Fields Here area in the spreadsheet. You should see the result shown in Figure E-15.

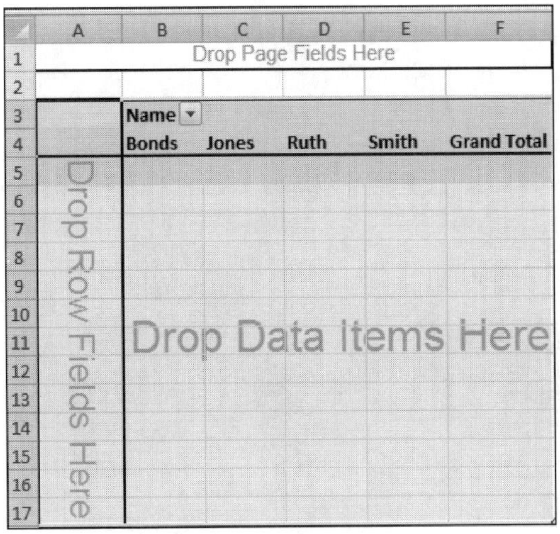

FIGURE E-15   Column fields

4. Next, take the Product field and drag it to the Drop Row Fields Here area. You should see the result shown in Figure E-16.

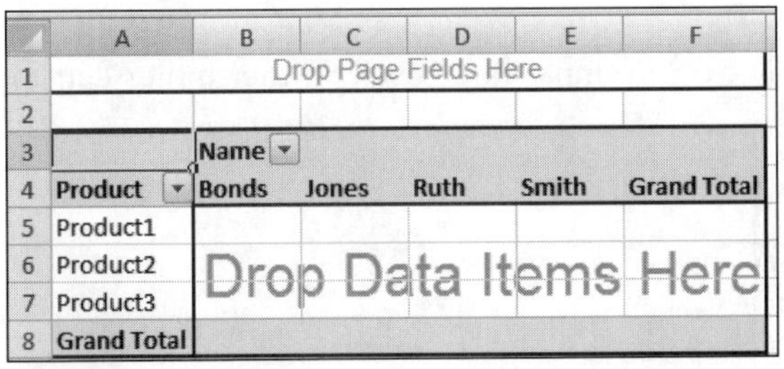

FIGURE E-16   Row fields

5. Finally, take the month fields (January, February, and March) and drag them individually to the Drop Data Items Here area to produce the finalized pivot table. You should see the result shown in Figure E-17.

	A	B	C	D	E	F	G
1	Drop Page Fields Here						
2							
3			Name				
4	Product	Values	Bonds	Jones	Ruth	Smith	Grand Total
5	Product1	Sum of January	25000	30000	44000	40000	139000
6		Sum of February	26000	35000	42000	38000	141000
7		Sum of March	25000	40000	33000	36000	134000
8	Product2	Sum of January	22000	33000	45000	41000	141000
9		Sum of February	25000	34000	40000	37000	136000
10		Sum of March	24000	45000	30000	38000	137000
11	Product3	Sum of January	19000	24000	50000	39000	132000
12		Sum of February	20000	30000	52000	50000	152000
13		Sum of March	19000	42000	35000	33000	129000
14	Total Sum of January		66000	87000	139000	120000	412000
15	Total Sum of February		71000	99000	134000	125000	429000
16	Total Sum of March		68000	127000	98000	107000	400000

FIGURE E-17 Data items

By default, Excel adds all of the sales for each salesperson by month for each product. At the bottom of the pivot table, Excel also shows the total sales for each month for all products.

BUILT-IN FUNCTIONS

The following functions are referred to in the Excel cases in this text:

MIN, MAX, AVERAGE, COUNTIF, and ROUND

The syntax of these functions is discussed here. The following examples are based on the runner data shown in Figure E-18.

	A	B	C	D	E	F	G
1	RUNNER#	LAST	FIRST	AGE	GENDER	HEIGHT	TIME (MIN)
2	100	HARRIS	JANE	O	F	60	70
3	101	HILL	GLENN	Y	M	65	70
4	102	GARCIA	PEDRO	M	M	76	85
5	103	HILBERT	DORIS	M	F	64	90
6	104	DOAKS	SALLY	Y	F	62	94
7	105	JONES	SUE	Y	F	64	95
8	106	SMITH	PETE	M	M	73	100
9	107	DOE	JANE	O	F	66	100
10	108	BRADY	PETE	O	M	73	100
11	109	BRADY	JOE	O	M	71	120
12	110	HEEBER	SALLY	M	F	59	125
13	111	DOLTZ	HAL	O	M	76	130
14	112	PEEBLES	AL	Y	M	76	63

FIGURE E-18 Runner data used to illustrate built-in functions

Note that the data is the same as that shown in Figure E-1, except that Figure E-18 has a column for the runner's height in inches.

MIN and MAX Functions

The MIN function determines the smallest value in a range of data. The MAX function returns the largest. Say that we want to know the fastest time for all runners, which would be the minimum time in column G. The MIN function computes the smallest value in a set of values. The set of values could be a data range or it could be a series of cell addresses separated by commas. The syntax of the MIN function is as follows:

MIN(set of data)

To show the minimum time in cell C16, you would enter the formula shown in Figure E-19's formula bar:

	C16		f_x	=MIN(G2:G14)		
	A	B	C	D	E	F	G
1	RUNNER#	LAST	FIRST	AGE	GENDER	HEIGHT	TIME (MIN)
2	100	HARRIS	JANE	O	F	60	70
3	101	HILL	GLENN	Y	M	65	70
4	102	GARCIA	PEDRO	M	M	76	85
5	103	HILBERT	DORIS	M	F	64	90
6	104	DOAKS	SALLY	Y	F	62	94
7	105	JONES	SUE	Y	F	64	95
8	106	SMITH	PETE	M	M	73	100
9	107	DOE	JANE	O	F	66	100
10	108	BRADY	PETE	O	M	73	100
11	109	BRADY	JOE	O	M	71	120
12	110	HEEBER	SALLY	M	F	59	125
13	111	DOLTZ	HAL	O	M	76	130
14	112	PEEBLES	AL	Y	M	76	63
15							
16	MINIMUM TIME:		63				

FIGURE E-19   MIN function in cell C16

(Assume you typed the label "MINIMUM TIME:" into cell A16.) You can see that the fastest time is 63 minutes.

To see the slowest time in cell G16, use the MAX function, whose syntax parallels the MIN function except that the largest value in the set is determined. See Figure E-20.

	G16		f_x	=MAX(G2:G14)		
	A	B	C	D	E	F	G
1	RUNNER#	LAST	FIRST	AGE	GENDER	HEIGHT	TIME (MIN)
2	100	HARRIS	JANE	O	F	60	70
3	101	HILL	GLENN	Y	M	65	70
4	102	GARCIA	PEDRO	M	M	76	85
5	103	HILBERT	DORIS	M	F	64	90
6	104	DOAKS	SALLY	Y	F	62	94
7	105	JONES	SUE	Y	F	64	95
8	106	SMITH	PETE	M	M	73	100
9	107	DOE	JANE	O	F	66	100
10	108	BRADY	PETE	O	M	73	100
11	109	BRADY	JOE	O	M	71	120
12	110	HEEBER	SALLY	M	F	59	125
13	111	DOLTZ	HAL	O	M	76	130
14	112	PEEBLES	AL	Y	M	76	63
15							
16	MINIMUM TIME:		63		MAXIMUM TIME:		130

FIGURE E-20   MAX function in cell G16

AVERAGE and ROUND Functions

The AVERAGE function computes the average of a set of values. Figure E-21 shows the use of the AVERAGE function in cell C-17:

	C17	▼		*fx*	=AVERAGE(G2:G14)		
	A	B	C	D	E	F	G
1	**RUNNER#**	**LAST**	**FIRST**	**AGE**	**GENDER**	**HEIGHT**	**TIME (MIN)**
2	100	HARRIS	JANE	O	F	60	70
3	101	HILL	GLENN	Y	M	65	70
4	102	GARCIA	PEDRO	M	M	76	85
5	103	HILBERT	DORIS	M	F	64	90
6	104	DOAKS	SALLY	Y	F	62	94
7	105	JONES	SUE	Y	F	64	95
8	106	SMITH	PETE	M	M	73	100
9	107	DOE	JANE	O	F	66	100
10	108	BRADY	PETE	O	M	73	100
11	109	BRADY	JOE	O	M	71	120
12	110	HEEBER	SALLY	M	F	59	125
13	111	DOLTZ	HAL	O	M	76	130
14	112	PEEBLES	AL	Y	M	76	63
15							
16	**MINIMUM TIME:**		63		**MAXIMUM TIME:**		130
17	**AVERAGE TIME:**		95.53846				

FIGURE E-21   AVERAGE function in cell C17

Notice that the value shown is a real number with many digits. What if you wanted to have the value rounded to a certain number of digits? Of course, you could format the output cell, but doing that only changes what is shown on the screen. You want the cell's contents actually to *be* the rounded number. Therefore, you need to use the ROUND function. Its syntax is:

ROUND(number, number of digits)

Figure E-22 shows the rounded average time (2 decimals) in cell G17.

	G17	▼		*fx*	=ROUND(C17,2)		
	A	B	C	D	E	F	G
1	**RUNNER#**	**LAST**	**FIRST**	**AGE**	**GENDER**	**HEIGHT**	**TIME (MIN)**
2	100	HARRIS	JANE	O	F	60	70
3	101	HILL	GLENN	Y	M	65	70
4	102	GARCIA	PEDRO	M	M	76	85
5	103	HILBERT	DORIS	M	F	64	90
6	104	DOAKS	SALLY	Y	F	62	94
7	105	JONES	SUE	Y	F	64	95
8	106	SMITH	PETE	M	M	73	100
9	107	DOE	JANE	O	F	66	100
10	108	BRADY	PETE	O	M	73	100
11	109	BRADY	JOE	O	M	71	120
12	110	HEEBER	SALLY	M	F	59	125
13	111	DOLTZ	HAL	O	M	76	130
14	112	PEEBLES	AL	Y	M	76	63
15							
16	**MINIMUM TIME:**		63		**MAXIMUM TIME:**		130
17	**AVERAGE TIME:**		95.53846		**ROUNDED AVERAGE:**		95.54

FIGURE E-22   ROUND function used in cell G17

To achieve this output, the cell C17 was used as the value to be rounded. Recall from Figure E-21 that cell C17 had the formula =AVERAGE(G2:G14). This ROUND formula would have given the same output in cell G17: =ROUND(AVERAGE(G2:G14),2). In this case, Excel evaluates the formula "inside out." First, the average function is evaluated, yielding the average with the many digits. That value is then input to the ROUND function and rounded to two decimals.

COUNTIF Function

The COUNTIF function counts the number of values in a range that meet a specified condition. The syntax is:

COUNTIF(range of data, condition)

The condition is a logical expression such as "=1", ">6", or "=F". The condition is shown with quotation marks, even if a number is involved.

Assume that you want to see the number of female runners in cell C18. Figure E-23 shows the formula used.

C18			f_x	=COUNTIF(E2:E14,"F")			
	A	B	C	D	E	F	G
1	RUNNER#	LAST	FIRST	AGE	GENDER	HEIGHT	TIME (MIN)
2	100	HARRIS	JANE	O	F	60	70
3	101	HILL	GLENN	Y	M	65	70
4	102	GARCIA	PEDRO	M	M	76	85
5	103	HILBERT	DORIS	M	F	64	90
6	104	DOAKS	SALLY	Y	F	62	94
7	105	JONES	SUE	Y	F	64	95
8	106	SMITH	PETE	M	M	73	100
9	107	DOE	JANE	O	F	66	100
10	108	BRADY	PETE	O	M	73	100
11	109	BRADY	JOE	O	M	71	120
12	110	HEEBER	SALLY	M	F	59	125
13	111	DOLTZ	HAL	O	M	76	130
14	112	PEEBLES	AL	Y	M	76	63
15							
16	MINIMUM TIME:		63		MAXIMUM TIME:		130
17	AVERAGE TIME:		95.53846		ROUNDED AVERAGE:		95.54
18	NUMBER OF FEMALES:		6				

FIGURE E-23   COUNTIF function used in cell C18

The logic of the formula is: Count the number of times that "F" appears in the data range E2:E14.

As another example of using COUNTIF, assume that column H shows the rounded ratio of the runner's height in inches to the runner's time in minutes (see Figure E-24).

H2			f_x	=ROUND(G2/F2,2)				
	A	B	C	D	E	F	G	H
1	RUNNER#	LAST	FIRST	AGE	GENDER	HEIGHT	TIME (MIN)	RATIO
2	100	HARRIS	JANE	O	F	60	70	1.17
3	101	HILL	GLENN	Y	M	65	70	1.08
4	102	GARCIA	PEDRO	M	M	76	85	1.12
5	103	HILBERT	DORIS	M	F	64	90	1.41
6	104	DOAKS	SALLY	Y	F	62	94	1.52
7	105	JONES	SUE	Y	F	64	95	1.48
8	106	SMITH	PETE	M	M	73	100	1.37
9	107	DOE	JANE	O	F	66	100	1.52
10	108	BRADY	PETE	O	M	73	100	1.37
11	109	BRADY	JOE	O	M	71	120	1.69
12	110	HEEBER	SALLY	M	F	59	125	2.12
13	111	DOLTZ	HAL	O	M	76	130	1.71
14	112	PEEBLES	AL	Y	M	76	63	0.83
15								
16	MINIMUM TIME:		63		MAXIMUM TIME:		130	
17	AVERAGE TIME:		95.53846		ROUNDED AVERAGE:		95.54	
18	NUMBER OF FEMALES:		6					

FIGURE E-24   Ratio of height to minutes in column H

Assume that all runners whose height in inches is less than their time in minutes will get an award. How many awards are needed? If the ratio is less than 1, an award is warranted. The COUNTIF function in cell G18 computes a count of ratios less than 1, as shown in Figure E-25.

	G18		f_x	=COUNTIF(H2:H14,"<1")			
	A	B	C	D	E	F	G	H
1	RUNNER#	LAST	FIRST	AGE	GENDER	HEIGHT	TIME (MIN)	RATIO
2	100	HARRIS	JANE	O	F	60	70	1.17
3	101	HILL	GLENN	Y	M	65	70	1.08
4	102	GARCIA	PEDRO	M	M	76	85	1.12
5	103	HILBERT	DORIS	M	F	64	90	1.41
6	104	DOAKS	SALLY	Y	F	62	94	1.52
7	105	JONES	SUE	Y	F	64	95	1.48
8	106	SMITH	PETE	M	M	73	100	1.37
9	107	DOE	JANE	O	F	66	100	1.52
10	108	BRADY	PETE	O	M	73	100	1.37
11	109	BRADY	JOE	O	M	71	120	1.69
12	110	HEEBER	SALLY	M	F	59	125	2.12
13	111	DOLTZ	HAL	O	M	76	130	1.71
14	112	PEEBLES	AL	Y	M	76	63	0.83
15								
16	MINIMUM TIME:		63		MAXIMUM TIME:		130	
17	AVERAGE TIME:		95.53846		ROUNDED AVERAGE:		95.54	
18	NUMBER OF FEMALES:		6		RATIOS<1:		1	

FIGURE E-25   COUNTIF function used in cell G18

PART 7

PRESENTATION SKILLS

TUTORIAL **F**

GIVING AN ORAL PRESENTATION

Giving an oral presentation provides you the opportunity to practice the presentation skills you'll need in the workplace. The presentations you create for the cases in this textbook will be similar to real-world presentations. You'll present objective, technical results to an organization's stakeholders, and you'll support your presentation with visual aids commonly used in the business world. During your presentation, your instructor might assign your classmates to role-play an audience of business managers, bankers, or employees and ask them to give you feedback on your presentation.

Follow these four steps to create an effective presentation:

1. Plan your presentation.
2. Draft your presentation.
3. Create graphics and other visual aids.
4. Practice your delivery.

You'll start at the beginning and look at the steps involved in planning your presentation.

PLANNING YOUR PRESENTATION

When planning an oral presentation, you need to be aware of your time limits, establish your purpose, analyze your audience, and gather information. This section will look at each of those elements.

Knowing Your Time Limits

You need to consider your time limits on two levels. First, consider how much time you'll have to deliver your presentation. For example, what can you expect to accomplish in ten minutes? The element of time is the driver of any presentation. It limits the breadth and depth of your talk—and the number of visual aids that you can use. Second, consider how much time you'll need for the actual process of preparing your presentation—that is, for drafting your presentation, creating graphics, and practicing your delivery.

Establishing Your Purpose

After considering your time limits, you must define your purpose: what you need and want to say and to whom you will say it. For the cases in the Access portion of this book, your purpose will be to inform and explain. For instance, a business's owners, managers, and employees may need to know how their organization's database is organized and how they could use it to fill in input forms and create reports. In contrast, for the cases in the Excel portion of the book, your purpose will be to recommend a course of action. You'll be making recommendations to business owners, managers, and bankers based on the results you obtained from inputting and running various scenarios.

Analyzing Your Audience

Once you have established the purpose of your presentation, you should analyze your audience. Ask yourself these questions: What does my audience already know about the subject? What do the audience members want to know? What do they need to know? Do they have any biases that I should consider? What level of technical detail is best suited to their level of knowledge and interest?

In some Access cases, you will make a presentation to an audience that might not be familiar with Access or with databases in general. In other cases, you might be giving a presentation to a business owner who started to work on the database but was not able to finish it. Tailor your presentation to suit your audience.

For the Excel cases, you will be interpreting results for an audience of bankers and business managers. The audience will not need to know the detailed technical aspects of how you generated your results. However, those listeners will need to know what assumptions you made prior to developing your spreadsheet, because those assumptions might have an impact on their opinion of your results.

Gathering Information

Because you will have just completed a case as you begin preparing your oral presentation, you'll have the basic information you need. For the Access cases, you should review the main points of the case and your goals. Make sure you include all of the points you think are important for the audience to understand. In addition, you might want to go beyond the requirements and explain additional ways in which the database could be used to benefit the organization now or in the future.

For the Excel cases, you can refer to the tutorials for assistance in interpreting the results from your spreadsheet analysis. For some cases, you might want to research the Internet for business trends or background information that you can use to support your presentation.

DRAFTING YOUR PRESENTATION

Now that you have completed the planning stage, you are ready to begin drafting your presentation. At this point, you might be tempted to write your presentation and then memorize it word for word. If you do, your presentation will sound unnatural because when people speak, they use a simpler vocabulary and shorter sentences than when they write. Thus, you might consider drafting your presentation by simply noting key phrases and statistics. When drafting your presentation, follow this sequence:

1. Write the main body of your presentation.
2. Write the introduction to your presentation.
3. Write the conclusion to your presentation.

Writing the Main Body

When you draft your presentation, write the body first. If you try to write the opening paragraph first, you'll spend an inordinate amount of time creating a "perfect" paragraph—only to revise it after you've written the body of your presentation.

Keeping Your Audience in Mind

To write the main body, review your purpose and your audience's profile. What are the main points you need to make? What are your audience's wants, needs, interests, and technical expertise? It's important to include some basic technical details in your presentation, but keep in mind the technical expertise of your audience.

What if your audience consists of people with different needs, interests, and levels of expertise? For example, in the Access cases, an employee might want to know how to input information into a form, but the business owner might already know how to input data and therefore be more interested in learning how to generate queries and reports. You'll need to acknowledge their differences in your presentation. For example, you might say, "And now, let's look at how data entry clerks can input data into the form."

Similarly, in the Excel cases, your audience will usually consist of business owners, managers, and bankers. The owners' and managers' concerns will be profitability and growth. In contrast, the bankers' main concern will be repayment of a loan. You'll need to address the interests of each group.

Using Transitions and Repetition

Because your audience can't read the text of your presentation, you'll need to use transitions to compensate. Words such as *next, first, second,* and *finally* will help your audience follow the sequence of your ideas. Words such as *however, in contrast, on the other hand,* and *similarly* will help the audience follow shifts in thought. You also can use your voice and hand gestures to convey emphasis.

Also think about how you can use body language to emphasize what you're saying. For instance, if you are stating three reasons, you can use your fingers to tick off each reason as you discuss it: one, two, three. Similarly, if you're saying that profits will be flat, you can make a level motion with your hand for emphasis.

As you draft your presentation, repeat key points to emphasize them. For example, suppose your point is that outsourcing labor will provide the greatest gains in net income. Begin by previewing that concept. State that you're going to demonstrate how outsourcing labor will yield the biggest profits. Then provide statistics that support your claim and show visual aids that graphically illustrate your point. Summarize by repeating your point: "As you can see, outsourcing labor does yield the biggest profits."

Relying on Graphics to Support Your Talk

As you write the main body, think of how you can best incorporate graphics into your presentation. Don't waste words describing what you're presenting when you can use a graphic to portray the subject quickly. For instance, instead of describing how information from a query is input into a report, show a sample, a query result, and a completed report. Figure F-1 and Figure F-2 show an Access query and the resulting report, respectively.

Customer Name	Customer City	Product Name	Quantity	Total Value
Altamar	Miami	Xlarge	15	$825.00
Café Pacific	Dallas	Medium	10	$300.00
Café Pacific	Dallas	Large	10	$420.00
Café Pacific	Dallas	Xlarge	10	$550.00
Red Lobster	Wichita	Xlarge	15	$825.00
Zee Grill	Toronto	Large	25	$1,050.00
Big Fish	Dearborn	Medium	5	$150.00
Big Fish	Dearborn	Large	5	$210.00
Big Fish	Dearborn	Xlarge	5	$275.00
The Park Hotel	Charlotte	Medium	3	$90.00
St. Elmo Steak House	Indianapolis	Large	10	$420.00
Bound'Ry Restaurant	Nashville	Xlarge	15	$825.00
Bellagio	Las Vegas	Large	5	$210.00
Osetra The Fish House	San Diego	Medium	12	$360.00

FIGURE F-1   Access query

Also consider what kinds of graphics media are available—and how well you can use them. For example, if you've never used Microsoft PowerPoint to prepare a presentation, will you have enough time to learn the software before you deliver your upcoming presentation? (If you don't know how to use PowerPoint, you might consider finding a tutorial on the Web to help you learn the basics.)

Anticipating the Unexpected

Even though you're just drafting your presentation at this stage, eventually you'll be answering questions from the audience. Being able to handle questions smoothly is the mark of a professional. The first steps to addressing audience questions are being able to anticipate them and preparing your answers.

You won't use all of the facts you gather for your presentation. However, as you draft your presentation, you might want to jot down some of those facts and keep them handy—just in case you need them to answer questions from the audience. For instance, during some Excel presentations, you might be asked why you are not recommending a certain course of action or why you did not mention that course of action in your report.

Writing the Introduction

After you have written the main body of your talk, you will want to develop an introduction. An introduction should be only a paragraph or two in length and should preview the main points that your presentation will cover.

For some of the Access cases, you might want to include general information about databases: what they can do, why they are used, and how they can help the company become more efficient and profitable. You won't need to say much about the business operation because the audience already works for the company.

For the Excel cases, you might want to include an introduction to the general business scenario and describe any assumptions you made when creating and running your decision support spreadsheet. Excel is used for decision support, so you should describe the choices and decision criteria you faced.

Writing the Conclusion

Every good presentation needs a good ending. Don't leave the audience hanging. Your conclusion should be brief—only a paragraph or two—and it should give your presentation a sense of closure. Use the conclusion to repeat your main points or, for the Excel cases, to recap your findings and/or recommendations.

Weekly Sales

Customer Name	Customer City	Product Name	Quantity	Total Value
Altamar	Miami			
		Xlarge	15	$825.00
Total Quantity and Value			15	$825.00
Bellagio	Las Vegas			
		Large	5	$210.00
Total Quantity and Value			5	$210.00
Big Fish	Dearbom			
		Xlarge	5	$275.00
		Large	5	$210.00
		Medium	5	$150.00
Total Quantity and Value			15	$635.00
Bound'Ry Restaurant	Nashville			
		Xlarge	15	$825.00
Total Quantity and Value			15	$825.00
Café Pacific	Dallas			
		Xlarge	10	$550.00
		Large	10	$420.00
		Medium	10	$300.00
Total Quantity and Value			30	$1,270.00
Osetra The Fish House	San Diego			

FIGURE F-2 Access report

CREATING GRAPHICS

Using visual aids is a powerful means of getting your point across and making it understandable to your audience. Visual aids come in a variety of forms, some of which are more effective than others.

Choosing Graphics Media

The media you use should depend on your situation and the media you have available. One of the key points to remember when using any media is this: *You must maintain control of the media, or you'll lose control of your audience.*

The following list highlights some of the most common media and their strengths and weaknesses:

- **Handouts:** This medium is readily available in classrooms and in businesses. It relieves the audience from taking notes. The graphics in handouts can be multicolored and of professional quality. *Negatives:* You must stop and take time to hand out individual pages. During your presentation, the audience may study and discuss your handouts rather than listen to you. Lack of media control is *the* major drawback—and it can hurt your presentation.
- **Chalkboard (or whiteboard):** This informal medium is readily available in the classroom but not in many businesses. *Negatives:* You need to turn your back to the audience when you write (thereby running the risk of losing the audience's attention), and you need to erase what you've written as you continue your presentation. With chalkboards, your handwriting must be good—that is, it should be legible even when you write quickly. In addition, attractive graphics are difficult to create.
- **Flip Chart:** This informal medium is readily available in many businesses. *Negatives:* The writing space is so small that it's effective only for a very small audience. This medium shares many of the same negatives as the chalkboard.

- **Overheads:** This medium is readily available in classrooms and in businesses. You have control over what the audience sees and when they see it. You can create professional PowerPoint presentations on overhead transparencies. *Negatives:* If you don't use an application such as PowerPoint, handwritten overheads look amateurish. Without special equipment, professional-looking graphics are difficult to prepare.
- **Slides:** This formal medium is readily available in many businesses and can be used in large rooms. You can use 35-mm slides or the more popular electronic on-screen slides. In fact, electronic on-screen slides are usually *the* medium of choice for large organizations and are generally preferred for formal presentations. *Negatives:* You must have access to the equipment needed for slide presentations and know how to use it. It takes time to learn how to create and use computer graphics. Also, you must have some source of ambient light; otherwise, you may have difficulty seeing your notes in the dark.

Creating Charts and Graphs

Technically, charts and graphs are not the same thing, although many graphs are referred to as charts. Usually charts show relationships and graphs show change. However, Excel makes no distinction and calls both entities charts.

Charts are easy to create in Excel. Unfortunately, they are so easy to create that people often use graphics that are meaningless or that inaccurately reflect the data the graphics represent. Next, you'll look at how to select the most appropriate graphics.

You should use pie charts to display data that is related to a whole. For example, you might use a pie chart when showing the percentage of shoppers who bought a generic brand of toothpaste versus a major brand, as shown in Figure F-3. (Note that when creating a pie chart, Excel takes the numbers you want to graph and makes them a percentage of 100.) You would *not*, however, use a pie chart to show a company's net income over a three-year period. While Figure F-4 does show such a pie chart, the graphic is not meaningful because it is not useful to think of the period as "a whole" or the years as its "parts."

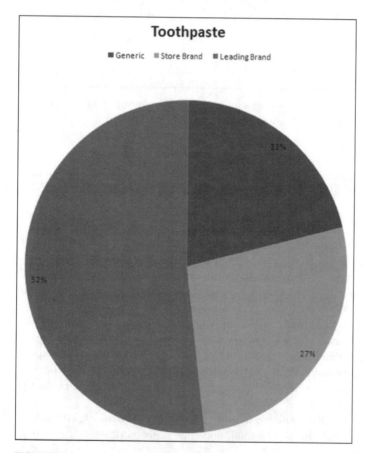

FIGURE F-3 Pie chart: appropriate use

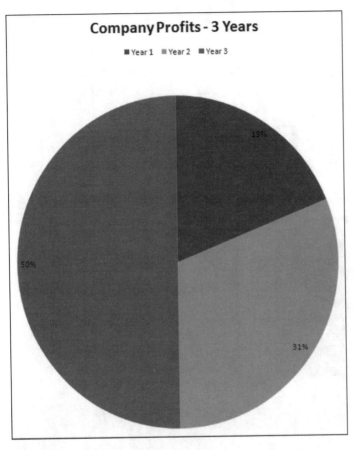

Company Profits - 3 Years

■ Year 1   ■ Year 2   ■ Year 3

19%

50%

31%

FIGURE F-4   Pie chart: inappropriate use

You should use bar charts when you want to compare several amounts at one time. For example, you might want to compare the net profit that would result from each of several different strategies. You also can use a bar chart to show changes over time. For example, you might show how one pricing strategy would increase profits year after year.

When you are showing a graphic, you need to include labels that explain what the graphic shows. For instance, when you're using a graph with an x- and y-axis, you should show what each axis represents so the audience doesn't puzzle over the graphic while you're speaking. Figure F-5 and Figure F-6 show the necessity of labels.

In Figure F-5, neither the graphic nor the x- and y-axes are labeled. Do the amounts shown correspond to units or dollars? What elements are represented by each bar? In contrast, Figure F-6 provides a comprehensive snapshot of the business operation, which would support a talk rather than distract from it.

Another common pitfall of producing visual aids is creating charts that have a misleading premise. For example, suppose you want to show how sales have increased and contributed to a growth in net income. If you simply graph the number of items sold in a given month, as displayed in Figure F-7, the visual may not give your audience any sense of the actual dollar value of those items. Therefore, it might be more appropriate (and more revealing) to graph the profit margin for the items sold times the number of items sold. Graphing the profit margin would give a more accurate picture of which item(s) are contributing to the increased net income. That graph is displayed in Figure F-8.

Something else you want to avoid is putting too much data in a single comparative chart. Here is an example: Assume you want to compare monthly mortgage payments for two loans with different interest rates and timeframes. You have a spreadsheet that computes the payment data, shown in Figure F-9.

In Excel, it is possible to capture all of that information in a single chart, such as the one shown in Figure F-10.

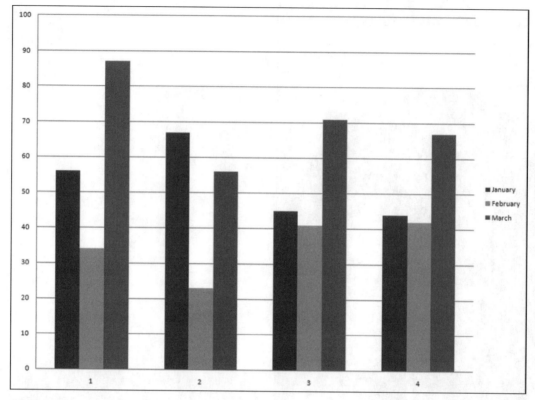

FIGURE F-5 Graphic without labels

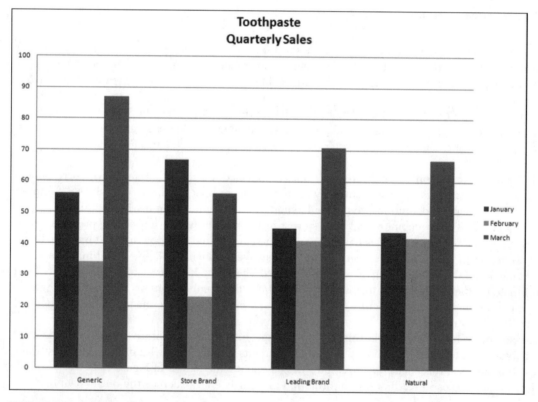

FIGURE F-6 Graphic with labels

Tutorial F

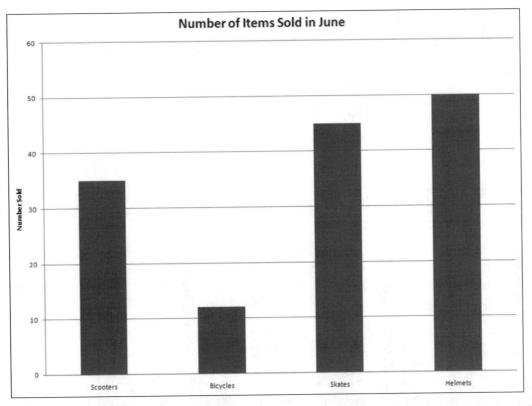

FIGURE F-7   Graph: number of items sold

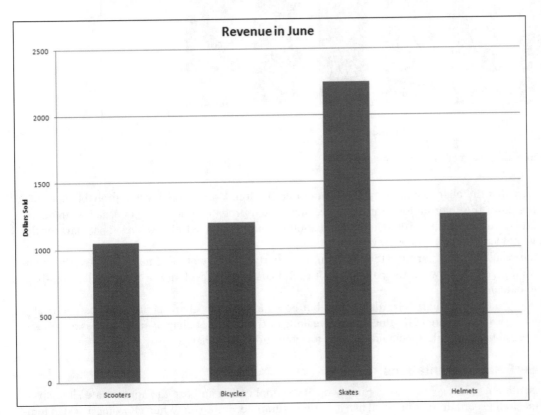

FIGURE F-8   Graph: profit of items sold

	A	B	C	D	E	F	G
1	**Calculation of Monthly Payment**						
2	Rate	6.00%	6.10%	6.20%	6.30%	6.40%	6.50%
3	Amount	$ 100,000	$ 100,000	$ 100,000	$ 100,000	$ 100,000	$ 100,000
4	Payment (360 payments)	$ 599	$ 605	$ 612	$ 618	$ 625	$ 632
5	Payment (180 payments)	$ 843	$ 849	$ 854	$ 860	$ 865	$ 871
6	Amount	$ 150,000	$ 150,000	$ 150,000	$ 150,000	$ 150,000	$ 150,000
7	Payment (360 payments)	$ 899	$ 908	$ 918	$ 928	$ 938	$ 948
8	Payment (180 payments)	$ 1,265	$ 1,273	$ 1,282	$ 1,290	$ 1,298	$ 1,306

FIGURE F-9 Calculation of monthly payment

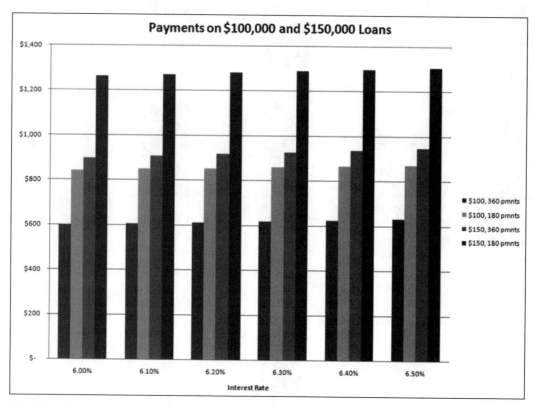

FIGURE F-10 Too much information in one chart

Note, however, that the chart contains a great deal of information. Most readers would probably appreciate your breaking down the information. For example, they would probably find the data easier to understand if you made one chart for the $100,000 loan and another chart for the $150,000 loan. The chart for the $100,000 loan would look like the one shown in Figure F-11.

A similar chart could be made for the $150,000 loan. The charts could then be augmented by text that summarizes the main differences between the payments for each loan. In that fashion, the reader is led step-by-step through the data analysis.

You might want to use the Chart Wizard in Excel, but be aware that the charting functions can be tricky to use, especially with sophisticated charting. Some tweaking of the resultant chart is often necessary. Your instructor might be able to provide specific directions for your individual charts.

Creating PowerPoint Presentations

PowerPoint presentations are easy to create. Simply open the application and use the appropriate slide layout for a title slide, a slide containing a bulleted list, a picture, a graphic, and so on. When choosing a design template (the background color, the font color and size, and the fill-in colors for all slides in your presentation), keep these guidelines in mind:

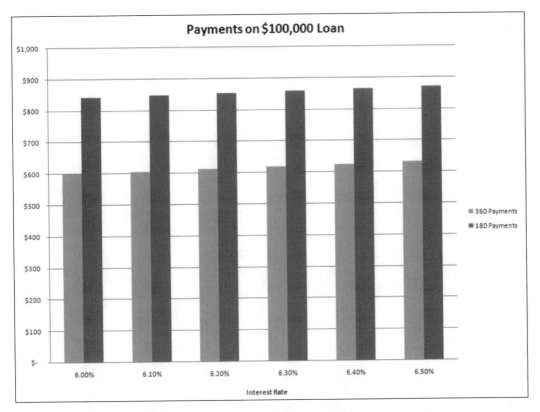

FIGURE F-11 Good balance of information

- Avoid using pastel background colors. Dark backgrounds such as blue, black, and purple work well on overhead projection systems.
- If your projection area is small or your audience is large, consider using boldfaced type for all of your text to make it more visible.
- Use transition slides to keep your talk lively. A variety of styles are available for use in PowerPoint. Common transitions include dissolves and wipes. Avoid wild transitions such as swirling letters; they will distract your audience from your presentation.
- Use Custom Animation effects when you do not want your audience to see the entire slide all at once. When you use an Entrance effect on each bullet point on a slide, the bullets come up one at a time when you click the mouse or the right arrow. This Custom Animation effect lets you control the visual aid and explain the elements as you go along. These types of Custom Animation effects can be incorporated and managed under the Custom Animation screen, as shown in Figure F-12.
- Consider creating PowerPoint slides that have a section for your notes. You can print the notes from the Print dialog box by choosing Notes Pages from the Print what drop-down menu, as shown in Figure F-13. Each slide is printed half size, with your notes appearing underneath each slide, as shown in Figure F-14.
- You should check your presentation on an overhead. What looks good on your computer screen might not be readable on an overhead screen.

Using Visual Aids Effectively

Make sure you've chosen the visual aids that will work most effectively. Also make sure you have enough—but not too many—visual aids. How many is too many? The amount of time you have to speak will determine the number of visual aids you should use, as will your audience. For example, if you will be addressing a group of teenage summer helpers, you might want to use more visual effects than if you were making a presentation to a board of directors. Remember to use visual aids to enhance your talk, not to replace it.

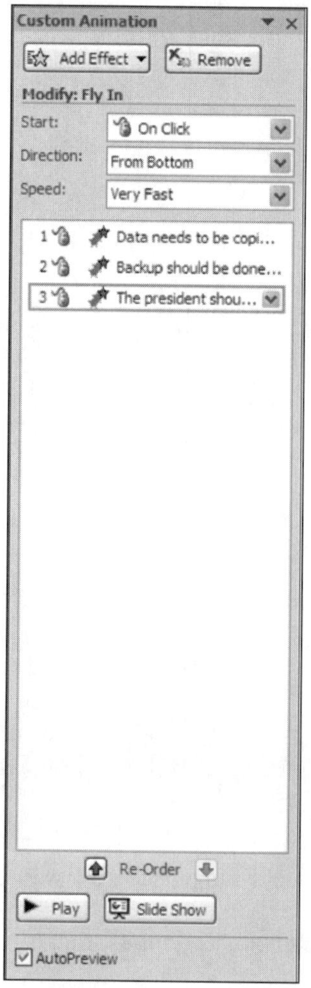

FIGURE F-12 Custom Animation screen

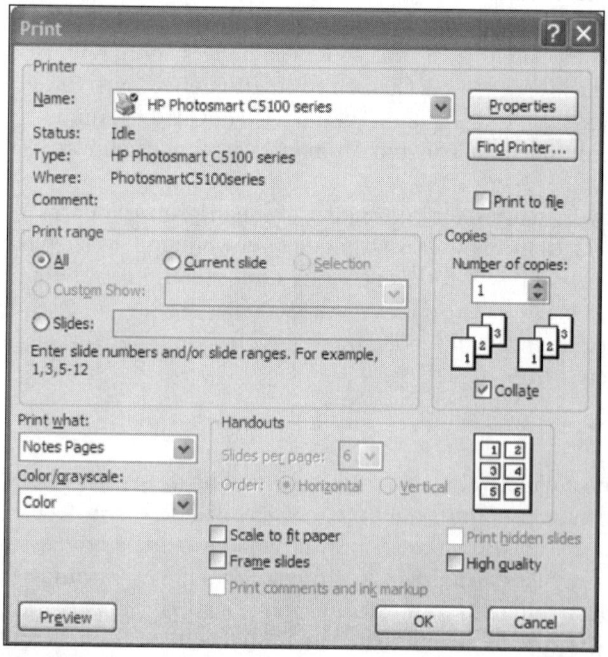

FIGURE F-13 Printing notes page

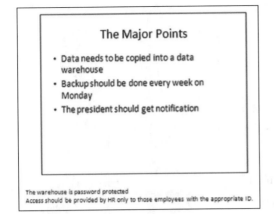

The Major Points

• Data needs to be copied into a data warehouse
• Backup should be done every week on Monday
• The president should get notification

The warehouse is password protected
Access should be provided by HR only to those employees with the appropriate ID.

FIGURE F-14 Sample notes page

Review each visual aid you've created to make sure it meets the following criteria:

- The size of the visual aid is large enough so that everyone in the audience can see it clearly and read any labels.
- The visual aid is accurate; for example, the graphics are not misleading, and there are no typos or misspelled words.
- The content of the visual aid is relevant to the key points of your presentation.
- The visual aid doesn't distract the audience from your message. Often when creating Power-Point slides, speakers get carried away with the visual effects; for example, they use spiraling text and other jarring effects. Keep your visuals professional-looking.
- A visual aid should look good in the presentation environment. If possible, try using your visual aid in the presentation environment before you actually deliver your presentation. For example, if you'll be using PowerPoint, try out your slides on an overhead projector in the room in which you'll be showing the slides. As previously mentioned, what looks good on your computer screen might not look good on the overhead projector when viewed from a distance of, say, 20 feet.
- All numbers should be rounded unless decimals or pennies are crucial.
- Slides should not look too busy or crowded. Most experts recommend that bulleted lists not contain more than four or five lines. You want to avoid using too many labels. For an example of a slide that is too busy and therefore likely to be ineffective, see Figure F-15.

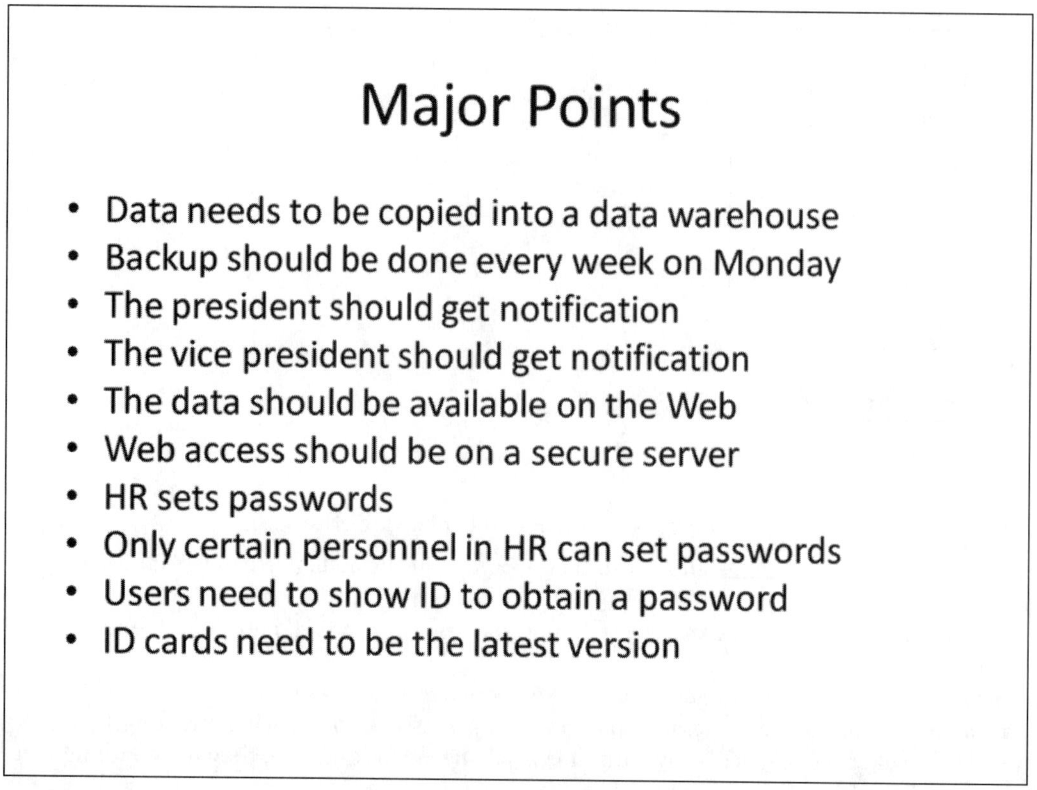

Major Points

- Data needs to be copied into a data warehouse
- Backup should be done every week on Monday
- The president should get notification
- The vice president should get notification
- The data should be available on the Web
- Web access should be on a secure server
- HR sets passwords
- Only certain personnel in HR can set passwords
- Users need to show ID to obtain a password
- ID cards need to be the latest version

FIGURE F-15   Busy slide

PRACTICING YOUR DELIVERY

Surveys indicate that for most people, public speaking is their greatest fear. However, fear or nervousness can be a positive factor. It can channel your energy into doing a good job. Remember that an audience is not likely to perceive you as being nervous unless you fidget or your voice cracks. Audience members want to hear the content of your talk, so think about them and their interests—not about how you feel.

The presentations you give for the cases in this textbook will be in a classroom setting with 20 to 40 students. Ask yourself this question: Am I afraid when I talk to just one or two of my classmates? The answer is probably no. Therefore, you should think of your presentation as an extended conversation with several of your classmates. Let your gaze shift from person to person and make eye contact with various people. As your gaze drifts around the room, say to yourself, I'm speaking to one person. As you become more experienced in speaking before a group, you will be able to let your gaze move naturally from one audience member to another.

Tips for Practicing Your Delivery

Giving an effective presentation is not—and should not—be like reading a report to an audience. Rather, it requires that you rehearse your message well enough so you can present it naturally and confidently and in tandem with well-chosen visual aids. Therefore, you must allow sufficient time to practice your delivery before you give your presentation. Here are some tips to help you hone the effectiveness of your delivery:

- Practice your presentation several times and use your visual aids when you practice.
- Show visual aids at the right time and only at the right time. A visual aid should not be shown too soon or too late. In your speaker's notes, you might include cues for when to show each visual aid.
- Maintain eye and voice contact with the audience when using the visual aid. Don't look at the screen or turn your back on the audience.
- Refer to your visual aids in your talk and with hand gestures. Don't ignore your own visual aid.
- Keep in mind that your visual aids should support your presentation, not *be* the presentation. In other words, don't include everything you are going to say on each slide. Use visual aids to illustrate key points and statistics and fill in the rest of the content with your talk.

- Check your time. Are you within the time limit?
- Use numbers effectively. When speaking, use rounded numbers; otherwise, you'll sound like a computer. Also make numbers as meaningful as possible. For example, instead of saying "in 84.7 percent of cases," say, "in five out of six cases."
- Don't "reach" to interpret the output of statistical modeling. For example, suppose you have input many variables into an Excel model. You might be able to point out a trend, but you might not be able to say with certainty that if a company employs the inputs in the same combination that you used them, the firm will get the same results.
- Record and then evaluate yourself. If that is not possible, have a friend listen to you and evaluate your style. Are you speaking down to your audience? Is your voice unnaturally high-pitched from fear? Are you speaking clearly and distinctly? Is your voice free of distractions, such as *um*, *you know*, *uh*, *so*, and *well*?
- If you use a pointer, either a laser pointer or a wand, be careful. Make sure you don't accidentally direct a laser pointer toward someone's face—you'll temporarily blind the person. If you're using a wand, don't swing it around or play with it.

Handling Questions

Fielding questions from an audience can be an unpredictable experience, because you can't anticipate all of the questions you might be asked. When answering questions from an audience, *treat everyone with courtesy and respect*. Use the following strategies to handle questions:

- Try to anticipate as many questions as possible and prepare answers in advance. Remember that you can gather much of the information to prepare those answers while you draft your presentation. Also, if you have a slide that illustrates a key point but doesn't quite fit in your presentation, save it; someone might have a question that the slide will answer.
- Mention at the beginning of the talk that you will take questions at the end of your presentation. That should prevent people from interrupting you. If someone tries to interrupt, smile and say that you'll be happy to answer all questions when you're finished or that the next graphic will answer the question. If, however, the person doing the interrupting is the CEO of your company, you should answer the question on the spot.
- When answering a question, repeat the question if you have *any* doubt that the entire audience might not have heard it. Then deliver the answer to the whole audience, not just the person who asked the question.
- Strive to be informative, not persuasive. In other words, use facts to answer questions. For instance, if someone asks your opinion about a given outcome, you might show an Excel slide that displays the Solver's output; then you can use that data as the basis for answering the question.
- If you don't know the answer to a question, don't try to fake it. For instance, suppose someone asks you a question about the Scenario Manager that you can't answer. Be honest. Say, "That is an excellent question; but unfortunately, I don't know the answer." For the classroom presentations you will be delivering as part of this course, you might ask your instructor whether he or she can answer the question. In a professional setting, you can say that you'll research the answer and e-mail the results to the person who asked the question.
- Signal when you are finished. You might say, "I have time for one more question." Wrap up the talk yourself.

Handling a "Problem" Audience

A "problem" audience or a heckler is every presenter's nightmare. Fortunately, such experiences are rare. If someone is rude to you or challenges you in a hostile manner, remain cool, be professional, and rely on facts. Know that the rest of the audience sympathizes with your plight and admires your self-control.

The problem you will most likely encounter is a question from an audience member who lacks technical expertise. For instance, suppose you explained how to input data into an Access form but someone didn't understand your explanation. In that instance, ask the questioner what part of the explanation was confusing. If you can answer the question briefly, do so. If your answer to the questioner begins to turn into a time-consuming dialogue, offer to give the person a one-on-one explanation after the presentation.

Another common problem is someone who asks you a question that you've already answered. The best solution is to answer the question as briefly as possible using different words (just in case the way in which you explained something confused the person). If the person persists in asking questions that have obvious answers, the person either is clueless or is trying to heckle you. In that case, you might ask the audience, "Who in the audience would like to answer that question?" The person asking the question should get the hint.

PRESENTATION TOOLKIT

You can use the form in Figure F-16 for preparation, the form in Figure F-17 for evaluation of Access presentations, and the form in Figure F-18 for evaluation of Excel presentations.

Preparation Checklist

Facilities and Equipment

☐ The room contains the equipment that I need.
☐ The equipment works and I've tested it with my visual aids.
☐ Outlets and electrical cords are available and sufficient.
☐ All the chairs are aligned so that everyone can see me and hear me.
☐ Everyone will be able to see my visual aids.
☐ The lights can be dimmed when/if needed.
☐ Sufficient light will be available so I can read my notes when the lights are dimmed.

Presentation Materials

☐ My notes are available, and I can read them while standing up.
☐ My visual aids are assembled in the order that I'll use them.
☐ A laser pointer or a wand will be available if needed.

Self

☐ I've practiced my delivery.
☐ I am comfortable with my presentation and visual aids.
☐ I am prepared to answer questions.
☐ I can dress appropriately for the situation.

FIGURE F-16 Preparation checklist

Evaluating Access Presentations

Course: _____ Speaker: _____ Date: _____

Rate the presentation by these criteria:
4=Outstanding   3=Good   2=Adequate   1=Needs Improvement
N/A=Not Applicable

Content

_____ The presentation contained a brief and effective introduction.

_____ Main ideas were easy to follow and understand.

_____ Explanation of database design was clear and logical.

_____ Explanation of using the form was easy to understand.

_____ Explanation of running the queries and their output was clear.

_____ Explanation of the report was clear, logical, and useful.

_____ Additional recommendations for database use were helpful.

_____ Visuals were appropriate for the audience and the task.

_____ Visuals were understandable, visible, and correct.

_____ The conclusion was satisfying and gave a sense of closure.

Delivery

_____ Was poised, confident, and in control of the audience

_____ Made eye contact

_____ Spoke clearly, distinctly, and naturally

_____ Avoided using slang and poor grammar

_____ Avoided distracting mannerisms

_____ Employed natural gestures

_____ Used visual aids with ease

_____ Was courteous and professional when answering questions

_____ Did not exceed time limit

Submitted by: _____

FIGURE F-17   Form for evaluation of Access presentations

Evaluating Excel Presentations

Course: _____ Speaker: _____ Date: _____

Rate the presentation by these criteria:
4=Outstanding   3=Good   2=Adequate   1=Needs Improvement
N/A=Not Applicable

Content

_____ The presentation contained a brief and effective introduction.

_____ The explanation of assumptions and goals was clear and logical.

_____ The explanation of software output was logically organized.

_____ The explanation of software output was thorough.

_____ Effective transitions linked main ideas.

_____ Solid facts supported final recommendations.

_____ Visuals were appropriate for the audience and the task.

_____ Visuals were understandable, visible, and correct.

_____ The conclusion was satisfying and gave a sense of closure.

Delivery

_____ Was poised, confident, and in control of the audience

_____ Made eye contact

_____ Spoke clearly, distinctly, and naturally

_____ Avoided using slang and poor grammar

_____ Avoided distracting mannerisms

_____ Employed natural gestures

_____ Used visual aids with ease

_____ Was courteous and professional when answering questions

_____ Did not exceed time limit

Submitted by: _____

FIGURE F-18   Form for evaluation of Excel presentations